Acute Topics in Sport Nutrition

Volume Editor

Manfred Lamprecht Graz

17 figures and 6 tables, 2012

Basel · Freiburg · Paris · London · New York · New Delhi · Bangkok ·
Beijing · Tokyo · Kuala Lumpur · Singapore · Sydney

Medicine and Sport Science

Founded 1968 by E. Jokl, Lexington, Ky.

Manfred Lamprecht, PhD, PhD
Centre for Physiological Medicine
Medical University of Graz
Harrachgasse 21/II
AT-8010 Graz (Austria), and
Green Beat – Institute of Nutrient Research and
Sport Nutrition
Petersbergenstrasse 95b
AT-8042 Graz (Austria)

Library of Congress Cataloging-in-Publication Data

Acute topics in sport nutrition / volume editor, Manfred Lamprecht.
    p. ; cm. -- (Medicine and sport science, ISSN 0254-5020 ; v. 59)
  Includes bibliographical references and indexes.
   ISBN 978-3-8055-9992-4 (hard cover : alk. paper) -- ISBN 978-3-8055-9993-1
(e-ISBN)
  I. Lamprecht, Manfred. II. Series: Medicine and sport science ; v. 59.
0254-5020
  [DNLM: 1. Dietary Supplements. 2. Nutritional Physiological Phenomena.
3. Sports--physiology. W1 ME649Q v.59 2012 / QU 145.5]

  613.7--dc23
                                    2012034260

Bibliographic Indices. This publication is listed in bibliographic services, including Current Contents®.

© Copyright 2012 by S. Karger AG, P.O. Box, CH–4009 Basel (Switzerland)
www.karger.com
Printed in Germany on acid-free and non-aging paper (ISO 9706) by Kraft Druck GmbH, Ettlingen
ISSN 0254–5020
e-ISSN 1662–2812
ISBN 978–3–8055–9992–4
e-ISBN 978–3–8055–9993–1

Contents

Preface

In many high-performance sports, far less than one percent difference in maximum power density decides between victory and defeat. An optimal diet and nutritional interventions can make the difference between winners and losers. Hence it is plausible that athletes and their carers have also realized the importance of optimal nutrition to achieve at least a minor advantage when it comes to competitive sports. In recent years, sport nutrition research has increased, and scientific journals, university courses, advanced education, scientific societies, etc., in the field have received more and more attention. The sport nutrition trade and related industries reacted and sales and marketing for sport nutrition products increased hand in hand with the interest of the consumers.

Many scientific publications exist about the classic themes of sport nutrition, like carbohydrate, protein or vitamin intake. On the other hand, scientific articles on topics that are not mainstream are still scarce although these issues are of importance; for example, more information is needed about specific sport supplements and dietary approaches to minimize the conflicting information circulating among carers and their athletes.

This volume of *Medicine and Sport Science* is devoted to 'Acute Topics in Sport Nutrition' and provides scientifically-based information with regard to the bioefficacy of trendy sport supplements and dietary approaches away from the mainstream. Experts in specific fields have attempted to inform and clarify under which circumstances the application of certain supplements and nutritional interventions could be beneficial, either for the performance or health of the athletes. At this juncture, I would like to thank all the contributors for taking the time to provide their expertise via excellent and scientifically-based articles.

The sixteen chapters of this book refer to a broad spectrum of current topics in sport nutrition: four chapters are dedicated to selected sport supplements off the mainstream claimed to influence lactate accumulation (β-alanine; R. Harris & C. Sale), blood flow (arginine, citrulline; A. Sureda & A. Pons), oxygen consumption and mitochondrial respiration (dietary nitrate; A.M. Jones, S.J. Bailey, A. Vanhatalo), and growth hormone response (γ-aminobutyric acid, M. Powers). The chapters about probiotics (M. Lamprecht & A. Frauwallner), immunoglukan (J. Majtan), bovine colostrum (G. Davison), fruit and vegetable concentrates (M. Lamprecht), cherry juice (K.S. Kuehl), and milk consumption plus resistance training (A.R. Josse & S.M. Phillips) refer to athletes' gut health, immune function, antioxidant potential, pain relief, and females' body composition and skeletal health. Hydration, hyperhydration and fluid balance/loading are covered by three chapters about glycerol use (S.P. van Rosendal & J.S. Coombes), salt and fluid loading (R. Mora-Rodriguez & N. Hamouti), and milk protein consumption (L. James). The chapters about chocolate milk (K. Pritchett & R. Pritchett) and l-carnitine (A. Huang & K. Owen) treat the important issue about exercise recovery. Finally, there is one special chapter – apart from certain ingredients, substances or dietary interventions – that informs about over-the-counter sport supplements and inadvertent doping (C. Judkins & P. Prock).

Sport nutrition advisors, sport physicians and scientists as well as coaches and interested athletes will benefit from the current information provided by this volume. The sport nutrition industry could draw benefits from the expert remarks in this book to create innovative ideas for the development of new and effective products. And it would also be desirable that this volume of *Medicine and Sport Science* stimulates research collaborations between sport nutrition scientists and companies in the field to improve the nutritional support to the target group that should benefit the most – the adult athlete.

Manfred Lamprecht
Medical University of Graz, Austria

Lamprecht M (ed): Acute Topics in Sport Nutrition.
Med Sport Sci. Basel, Karger, 2013, vol 59, pp 1–17

Beta-Alanine Supplementation in High-Intensity Exercise

Roger C. Harris[a] · Craig Sale[b]

[a]Junipa Ltd, Newmarket, Suffolk, and [b]Sport, Health and Performance Enhancement (SHAPE) Research Group, School of Science and Technology, Nottingham Trent University, Nottingham, Notts., UK

Abstract

Glycolysis involves the oxidation of two neutral hydroxyl groups on each glycosyl (or glucosyl) unit metabolised, yielding two carboxylic acid groups. During low-intensity exercise these, along with the remainder of the carbon skeleton, are further oxidised to CO_2 and water. But during high-intensity exercise a major portion (and where blood flow is impaired, then most) is accumulated as lactate anions and H^+. The accumulation of H^+ has deleterious effects on muscle function, ultimately impairing force production and contributing to fatigue. Regulation of intracellular pH is achieved over time by export of H^+ out of the muscle, although physicochemical buffers in the muscle provide the first line of defence against H^+ accumulation. In order to be effective during high-intensity exercise, buffers need to be present in high concentrations in muscle and have pK_as within the intracellular exercise pH transit range. Carnosine (β-alanyl-L-histidine) is ideal for this role given that it occurs in millimolar concentrations within the skeletal muscle and has a pK_a of 6.83. Carnosine is a cytoplasmic dipeptide formed by bonding histidine and β-alanine in a reaction catalysed by carnosine synthase, although it is the availability of β-alanine, obtained in small amounts from hepatic synthesis and potentially in greater amounts from the diet that is limiting to synthesis. Increasing muscle carnosine through increased dietary intake of β-alanine will increase the intracellular buffering capacity, which in turn might be expected to increase high-intensity exercise capacity and performance where this is pH limited. In this study we review the role of muscle carnosine as an H^+ buffer, the regulation of muscle carnosine by β-alanine, and the available evidence relating to the effects of β-alanine supplementation on muscle carnosine synthesis and the subsequent effects of this on high-intensity exercise capacity and performance.

Dedication

This paper is dedicated to the memory of Dr. John Wise, of Natural Alternatives International, San Marcos, California, USA, who was initially invited by the editor to be a co-author of this review, but who sadly passed away in November 2011. Dr. Wise played a key role in the early investigations of the effects of β-alanine supplementation on muscle carnosine synthesis and performance.

$CCT_{110\%}$	cycle capacity test and 100% of power max
b.w.	body weight
dm	dry muscle
E-C	excitation-contraction coupling
G-protein	guanine nucleotide-binding proteins
^{1}H-MRS	proton magnetic resonance spectroscopy
HCD	histidine-containing dipeptide
K_m	Michaelis constant (the substrate concentration supporting a reaction rate half of maximum, in an enzyme-catalysed reaction)
Lac^-	lactate anion
H^+	hydrogen cation
M-Carn	muscle carnosine
Mrg	Mas-related gene receptors (Mrgs form a large family of G-protein-coupled receptors expressed in dorsal root ganglia and trigeminal ganglia associated with sensory neurons that detect painful stimuli)
MVIC	maximal voluntary isometric contraction
PCr	phosphorylcreatine
pH_i	intracellular pH
pK_a	acid dissociation constant
SR	sustained release
$t_{1/2}$	half-life
Td_1	defines time-delay 1 for the first monoexponential term computing respiratory oxygen uptake
$Vo_{2\ max}$	maximal oxygen uptake

Background

β-Alanine is found in muscle in combination with L-histidine forming the dipeptide, carnosine (β-alanyl-L-histidine, abbreviated in the context of the muscle content as M-Carn). Carnosine is a member of a family of three related HCDs, the others being anserine (β-alanyl-L-(1-methyl)-histidine) and balenine (β-alanyl-L-(3-methyl)-histidine. Carnosine and related HCDs are found in high concentrations in skeletal muscle of both vertebrates and non-vertebrates, as well as in the central nervous system.

Carnosine was first measured in human muscle by Bergstrom et al. [1] where it is the only HCD present. In human muscle, for instance m. vastus lateralis, the molar concentration of M-Carn varies from 4 to 20 mM (12–60 mmol·kg^{-1} dry muscle) making it one of most abundant small molecular weight compounds present in resting muscle after PCr (75 mmol·kg^{-1} dm), creatine (49 mmol·kg^{-1} dm) and ATP (24 mmol·kg^{-1} dm) [2]. Factors determining the concentration of M-Carn in humans include muscle fibre composition, with a 1.3–2 times higher concentration in type II compared to type I muscle fibres [3–6] and dietary intake of β-alanine [7]. Age and gender may also influence the carnosine content in muscle [8, 9] although the evidence for this is somewhat tenuous, at least in humans, as the changes reported could be secondary to changes in fibre composition (including changes in mean fibre area, as a determinant of the volume occupied by each fibre type in the muscle sampled) and diet [7]. Suzuki et al. [10] reported a doubling in M-Carn with 8 weeks of training, consisting of 28 bouts of multiple 30-s Wingate tests performed on a cycle ergometer (a total exercise time over the 8 weeks of 14 min). Other studies using 4–16 weeks of intensive sprint training [6, 11], 12 weeks of whole-body training [12] and 4 weeks of unilateral cycle training [13] have, however, shown no effect on M-Carn. Chronic training, however, is associated with an increase in M-Carn [5, 14], although again this may be secondary to changes in fibre composition (and mean fibre area) and diet.

Role of M-Carn as Buffer of H^+ during High-Intensity Exercise

High-intensity exercise is associated with the formation of two carboxylic acid groups arising from the oxidation of neutral hydroxyl groups within each glucose or glycosyl (from glycogen) unit metabolised. Both carboxylic acid groups are fully dissociated over the physiological pH range, with most accumulated as Lac^- and H^+ in muscle. In reality, H^+ exist mostly in the hydrated state, bound to one or more molecules of water for instance as the hydronium ion (H_3O^+). However, for convenience the term H^+ is used to define all such states.

Lac^- production accounts for the generation of 94% of the H^+ produced in skeletal muscle during high-intensity exercise [15] causing a decline in intracellular pH (pH_i) from around 7.0 at rest [16] to as low as 6.0 [17]. H^+ accumulation may contribute to fatigue by interfering with several metabolic processes affecting force production [18]. More specifically, the accumulation of H^+ in skeletal muscle disrupts the recovery of PCr [19] and its role as a temporal buffer of ADP accumulation [20–22], inhibits glycolysis [23] and disrupts functioning of the muscle contractile machinery [24, 25]. Regulation of pH_i is achieved in the long term by export of H^+, but in the short term by a limited range of intracellular buffers which include carnosine and its methylated derivatives.

To be effective, buffers need to occur in high concentrations and to have a pK_a within the exercise pH_i transit range. Carnosine satisfies both requirements occurring in the low millimolar range in human muscle while the imidazole ring of the histidine residue exhibits a pK_a of 6.83 [26]. Other potential buffers in skeletal muscle with pK_as in or close to the pH_i transit range include proteins, inorganic and organic phosphates, and bicarbonate present in cells at the start of contraction. The contribution of proteins is limited to their respective histidine contents as no other amino acid has a side chain with an effective pK_a. Organic phosphates include nucleotide phosphates and hexose phosphates, principally glucose 6-phosphate, fructose 6-phosphate, and glycerol 1-phosphate, which accumulate with an increased glycolytic rate during intense exercise. PCr, one of the largest metabolic pools of phosphate, has a pK_a of

4.58 and does not itself contribute to buffering in the resting state. However, with the commencement of exercise, net decline in the PCr pool, matched by increases in the concentration of inorganic and organic phosphates, will increase its contribution. As a result of the changes in the PCr pool, the physicochemical buffering capacity of muscle increases with exercise, reaching a theoretical peak when all PCr has been hydrolysed.

Estimates of the importance of M-Carn to intracellular physicochemical buffering determined by acid titration of muscle homogenates [27] or by calculation [15] both indicate a contribution in human skeletal muscle (with an M-Carn concentration of ~20 mmol·kg^{-1} dm) of between 5 and 10%. However, implicit in these estimates (as a result of the methods used) is the assumption that during exercise, (1) all muscle-located proteins, both intracellular and extracellular, are available to contribute to H$^+$ buffering, (2) that hydrolysis of PCr is complete, and (3) that the distribution of carnosine in muscle is uniform. Point 1 is incorrect and will result in an overestimation of the contribution of non-carnosine buffers, lessening the apparent relative importance of M-Carn. In the absence of modelling for the partial hydrolysis of PCr during muscle contraction, point 2 will similarly overestimate non-carnosine buffering. With respect to point 3, the distribution of carnosine in human muscle is not uniform with the content in fast-twitch fibres being 1.3–2 times higher than in slow twitch [3, 4, 6, 28]. This will underestimate the importance of carnosine in type II fibres where its role as a pH buffer would be expected to be greatest; though equally it would overestimate its importance in type I muscle fibres.

The combination of β-alanine, a non-proteogenic amino acid (not to be confused with α-alanine found in most proteins), with histidine raises the pK$_a$ of the imidazole ring from pH ~6.0 to ~6.8, improving its effectiveness in buffering H$^+$ over the exercise pH$_i$ transit range. More importantly, combination with β-alanine renders the histidine inert to participation in proteogenesis, enabling high concentrations to be accumulated in muscle cells. This affords a more efficient means to vary the intracellular physicochemical buffering capacity than by alteration of the protein content, where histidine is only 1 of 20 amino acids. When considered along with the other HCDs, this has provided a highly efficient means for species to vary the intracellular buffering capacity of muscle by evolution, matching the exercise demands imposed by escape, combat or hunting in the wild.

Other Suggested Physiological Roles of M-Carn

Other physiological roles have been ascribed to carnosine in muscle; including protection of proteins against glycation by acting as a sacrificial peptide [29, 30], the prevention of protein-protein cross-links through reactions with protein-carbonyl groups [29, 31], acting as an antioxidant (for reviews, see Boldyrev et al. [32] and Boldyrev [33]) and increasing calcium sensitivity in muscle fibres augmenting force production and total work done [34–38]. However, few of the ascribed physiological roles for carnosine, other than its role as a pH buffer, have been shown to occur in vivo and in humans. Indeed the majority of the work cited above has been conducted in vitro.

Arguably, the potentiation of calcium sensitivity of E-C coupling in both type I and II human muscle fibres could have relevance to muscle function in vivo but needs further characterisation in terms of field performance (strength or endurance). Lamont and Miller [34] showed that the presence of carnosine reduced the amount of Ca^{2+} ions required to produce half-maximum tension in chemically skinned cardiac and skeletal muscle. They reported an increase in maximal force production by different muscle types and suggested that the higher concentrations of carnosine shown in fast-twitch fibres might relate to an effect of enhanced Ca^{2+} ion sensitivity. Dutka and Lamb [37] and Dutka et al. [38] showed that the increase in Ca^{2+} ion sensitivity on E-C coupling occurred in a concentration-dependent manner and suggested that higher concentrations of M-Carn could, by this mechanism, delay the onset of fatigue. However, it is by no means certain that Ca^{2+} ion sensitivity, and changes in this, are factors integral to fatigue in whole-body exercise[1].

Notwithstanding any additional physiological role, M-Carn and the HCDs in general remain important contributors to pH_i regulation in muscle and in practice the only means to vary the buffering capacity within and across species.

Regulation of M-Carn by β-Alanine

Carnosine in muscle is synthesised in situ by the action of carnosine synthase

$$ATP + \text{L-histidine} + β\text{-alanine} \rightarrow AMP + \text{diphosphate} + \text{carnosine}$$

Synthesis of carnosine appears limited by the low concentration in muscle cells of β-alanine in comparison to the high K_m (1.0–2.3 mM) that β-alanine has for carnosine synthase [40, 41]. In contrast, histidine is present in much higher concentrations in muscle, has a much lower K_m (16.8 μM) for the synthase [42], and is unlikely to be limiting to carnosine synthesis. β-Alanine is synthesised in the liver as the final metabolite of uracil and thymine degradation [43, 44] before being transported via the blood to muscle and other tissues. To this will be added β-alanine available from carnosine and other HCDs absorbed with the ingestion of muscle meat [45–48], and

---

[1] Whereas there is a rationale for pH_i decrease causing muscle fatigue, and there is extensive evidence from laboratory studies in support of this, again there is a lack of evidence at the level of whole-body exercise. A possible exception to this is the study of Hultman et al. [39] where ingestion of ammonium chloride (0.3 g·kg^{-1} b.w.) was used to induce a moderate acidosis prior to in situ percutaneous electrical stimulation of muscle. As a result, pH_i was lower at the end of 75 s stimulation (pH 6.54 as opposed to 6.70) as was also the final force sustained (44.6% of initial compared with 55.4% without ammonium chloride.)

hydrolysed to their constituent amino acids by the action of carnosinase present in intestinal mucosa and serum [49–51]. The transport of β-alanine into muscle is mediated by a specific β-amino acid transport protein that is dependent upon stoichiometric concentrations of both Na^+ and Cl^- in a 2:1:1 (Na^+:Cl^-:β-amino acid) ratio [52, 53] and exhibits a K_m of ~40 μM with respect to β-alanine [54].

In vegetarians, M-Carn is limited by hepatic synthesis of β-alanine resulting in comparatively low contents of the order of 13 mmol·kg^{-1} dm determined in muscle biopsies of m. vastus lateralis by HPLC from a subject group mostly comprising aerobically trained female UK athletes [55, 56]. Similarly, a reduced level of M-Carn in vegetarians was reported by Everaert et al. [57] with reductions of 17–26% seen in m. soleus, m. gastrocnemius and m. tibialis anterior, compared with Belgian subjects eating a mixed diet.

In omnivores, de novo hepatic β-alanine synthesis may be augmented by the hydrolysis of dietary-supplied HCDs from muscle meat resulting in M-Carn levels two or more times higher than in vegetarians [55]. Hydrolysis of dietary-supplied HCD from, for instance, the ingestion of chicken broth, has been shown to supply close to the theoretical level of the amount of β-alanine present in the bound HCD form [48]. Direct supplementation of the diet with β-alanine over a period of weeks will similarly increase M-Carn. In the first of a series of studies [48] where multiple doses of 800 mg β-alanine were given per day over 4 weeks in gelatine capsules, the mean increase in M-Carn in m. vastus lateralis was 60%. When extended to 10 weeks, the increase was 80% with absolute values now close to 40 mmol·kg^{-1} dm [4]. A maximal single dose of 800 mg, on average 10 mg·kg^{-1} b.w., was used in these studies to avoid symptoms of paraesthesia, and is equivalent to the amount of β-alanine available from the ingestion of 150 g of chicken breast meat, assuming hydrolysis of the HCDs present. A maximum of 8 such doses was given in a single day without any negative effects such as increased feelings of paraesthesia, clinical chemistry or ECG. Supplementation with L-carnosine itself resulted in the same increase in muscle over 4 weeks of supplementation as an isomolar dose of β-alanine, showing no additional effect of the histidine also released on hydrolysis [48].

With curtailment of supplementation, M-Carn returns to the presupplementation level in m. vastus lateralis with an estimated $t_{1/2}$ of ~9 weeks [58, 59]. Estimates of $t_{1/2}$ following elevation of M-Carn in m. anterior tibialis, m. gastrocnemius and m. soleus range from 5 to 8 weeks [59–61]. No information is available on the rates of decline of M-Carn specifically in types I and II fibres; a faster rate of decline could possibly explain the lower M-Carn content observed in type I fibres. The mechanism governing the loss of M-Carn accumulated during supplementation is unknown, but the possibilities are a slow release from muscle fibres or destruction within the fibres due to reaction with free radicals or carbonyl groups [62, 63].

In equines there is no measureable loss of M-Carn with acute exercise [64]. However, exercise-induced muscle damage may result in temporarily raised plasma concentrations in equines where plasma carnosinase is absent [65] but the increases are of an order which would be barely measureable as a change in M-Carn in muscle. No comparable measurements have been performed in humans, although it is clear that there is no progressive loss of M-Carn with chronic training [6, 11].

Single doses of β-alanine in excess of 10 mg·kg^{-1} b.w., or the equivalent molar dose of L-canosine, administered in solution as a drink or in gelatine capsules cause symptoms of flushing together with a prickly sensation affecting (in approximate order of occurrence) the face and ears, neck and shoulders, arms, hands and upper trunk, and lower trunk [48]. Symptoms, also termed paraesthesia, last 15–60 min and in terms of severity appear dose-dependent [66]. At 10 mg·kg^{-1} b.w., only very mild symptoms may be experienced by a small percentage of subjects but at 40 mg·kg^{-1} b.w. severe and uncomfortable symptoms are experienced by most.

Several possible mechanisms exist to account for the symptoms of paraesthesia, including β-alanine activation of strychnine-sensitive glycine receptor sites, associated with glutamate sensitive N-methyl-D-aspartate receptors in the brain and central nervous system [67–69] and activation of the Mrg (mas-related gene) family of G-protein-coupled receptors, which are triggered by interactions with specific ligands, such as β-alanine [70]. MrgD-containing dorsal route ganglia neurons terminate in the skin, but not in blood vessels, muscle, or other major internal organs and participate in the modulation of neuropathic pain. To paraphrase Crozier et al. [70], 'neuropathic pain is qualitatively different from ordinary pain and is usually perceived as a steady burning, pins and needles and/or electric shock sensation'. This is an accurate description of the symptoms experienced by subjects reporting sensations of paraesthesia after taking high doses of β-alanine or L-carnosine.

To circumvent symptoms of paraesthesia, early β-alanine supplementation studies used a maximum single dose of 800 mg, corresponding to ~10 mg·kg^{-1} b.w., and administered up to 8 times a day to give a total dose of 6.4 g [4, 48]. More recent supplementation studies have used a sustained release formulation (CarnoSyn™ SR, from Natural Alternatives International, San Marcos, Calif., USA, available to the public as High Intensity™, from Power Bar, Florham Park, N.J., USA) enabling two 800-mg SR tablets to be given simultaneously without symptoms of paraesthesia [59, 71–74].

Ergogenic Effect of Raised M-Carn

As reviewed by Hobson et al. [75], a number of studies have demonstrated a significant effect of β-alanine-induced M-Carn elevation on exercise performance. In this meta-analysis, 15 published peer-reviewed studies were assessed, from which it was concluded that exercise lasting 60–240 s was improved (p = 0.001) with β-alanine supplementation, as was exercise of >240 s (p = 0.046). In contrast, there was no benefit of β-alanine supplementation on sprint exercise lasting <60 s (p = 0.312), consistent with the mechanism for the improvement in performance being linked to an increase in H+ buffering capacity, as opposed to an effect of increased Ca^{2+} ion sensitivity on E-C coupling.

The median effect of β-alanine supplementation on exercise >60 s was a 2.85% (range −0.37 to 10.49%) improvement in exercise capacity. Of particular note was the study of Hill et al. [4] which used a $CCT_{110\%}$ output with an expected endurance time of 150 s, where 4 weeks of β-alanine supplementation resulted in a ~60% increase in M-Carn and a 11.8% improvement in endurance time. This was subsequently repeated by Sale et al. [71], using the same exercise protocol, when an increase of 12.1% in endurance time was recorded. Evidence that the effect of β-alanine supplementation was due to an increase in H+ buffering capacity was suggested by a positive gain in performance (+6.5%) when subjects were supplemented with sodium bicarbonate (0.3 g·kg^{-1} b.w.), and with the combination of β-alanine and sodium bicarbonate resulting in a 16.2% increase in cycling capacity.

In three studies where both performance measures have been made along with measurements of M-Carn, improvement in exercise capacity was positively correlated with the increase in M-Carn. The exercise modalities included cycling with an expected pre-supplementation endurance time of 150 s [4], rowing over a distance of 2,000 m [76], and time-to-exhaustion in a constant-load submaximal test and incremental test in 60- to 80-year-old subjects [72]. The demonstration of a positive cause-and-effect relationship between increased M-Carn and performance, in these three studies, is almost unique in studies of dietary supplements.

A positive effect of β-alanine supplementation on isometric endurance of the knee extensors contracting at 45% of MVIC force has recently been reported by Sale et al. [77]. Lac^{-} plus pyruvate accumulation by the time of fatigue varies with the % MVIC sustained, reaching a peak at 45% MVIC [78]. As the circulation to the contracting muscles is largely occluded at this intensity by the increase in intramuscular pressure, Lac^{-} and H+ loss from the contracting muscle is minimal. Based upon this, the expected endurance time at 45% MVIC is around 78 s [78], with baseline data from Sale et al. [77] being within 3–4 s of this. Endurance time after 4 weeks of β-alanine supplementation at 6.4 g·day^{-1} was increased by a mean of 9.7 s (13.2%) with the estimated increase in H+ production (estimated from the known rate of Lac^{-} plus pyruvate production at this intensity [78] closely matched to the increase in H+ buffering capacity from the estimated increase in M-Carn (estimated from the changes observed in Hill et al. [4] and Kendrick et al. [6]). Whilst isometric exercise is involved in a number of sports (e.g., rock climbing and dingy sailing), it is more often encountered in everyday manual activities involving the lifting and carrying of heavy weights.

Effect on Maximal/Supramaximal Exercise

Since completion of the meta-analysis by Hobson et al. [75], three further studies have been published on the effects of β-alanine supplementation on maximal/

supramaximal exercise. Saunders et al. [73] showed no effect on repeated sprint ability using the Loughborough Intermittent Shuttle Test. Although over a longer duration that the 60 s limit suggested by Hobson et al. [75], but still within the category of supramaximal sprint exercise, Jagim et al. [79] also showed no effect of β-alanine supplementation on sprint exercise at 115% (lasting ~140–165 s) and 140% $Vo_{2\ max}$ (lasting ~65–75 s).

From these studies it would seem that β-alanine supplementation has little effect on high-intensity exercise of short duration. However, in contrast to these studies, van Thienen et al. [80] observed a significant effect of supplementation on a 30-s cycle sprint when this followed a 110-min simulated cycle race and a 10-min time trial. Under these conditions the sprint exercise would most likely have begun after muscle pH_i had been lowered by the preceding exercise, and this may explain why supplementation was effective in this case. If so, this could be highly significant for sports where a sprint finish is called for following a sustained bout of exercise, such as indeed may occur in cycle racing. Similarly, a significant effect of β-alanine supplementation was recorded by Saunders et al. [74] for the YoYo protocol, which is considered relevant to the type of activity undertaken during football (and potentially other invasion games such as rugby, hockey and netball). In the study of Saunders et al. [74], 17 amateur male football players undertook repeated bouts of 2×20-m runs against a target time, with 10 s active recovery between bouts. Sprint bouts were maintained until subjects failed to reach the finish line within the allotted time on two consecutive occasions; the total distance covered being used to define performance. Nine subjects were supplemented with 3.2 g·day^{-1} β-alanine in the form of CarnoSyn^{TM} SR tablets for 12 weeks, and 8 with a matched placebo. β-Alanine supplementation resulted in a significant increase (34%) in sprint distance covered compared to a (within-group, non-significant) 7.3% decline in those subjects supplemented with placebo.

Effect on Longer Duration Submaximal Exercise

Hobson et al. [75] concluded that β-alanine supplementation had a highly significant effect on exercise of >60 s duration. Their conclusion was based upon the results from 13 studies. Included in these were 4 studies which examined the effect of supplementation on the physical working capacity at the neuromuscular fatigue threshold [81–84]. These indicated that supplementation may exhibit effects long before attainment of fatigue, as shown by delayed changes in electromyography denoting the neuromuscular fatigue threshold, and leading to increases in the physical working capacity. Such effects were observed in males and females and in young and elderly. Supplemented with β-alanine for 90 days at 2.4 g·day^{-1}, 12 elderly (aged 55–92 years) men and women showed a 29% increase in the physical working capacity at the neuromuscular fatigue threshold, whereas comparable subjects treated with placebo showed no change during the same period [83]. This was the first study conducted in the elderly but lacked direct evidence of the effectiveness of β-alanine in raising M-Carn.

Proof of the effectiveness of dietary β-alanine on M-Carn in the elderly has recently been provided by Favero et al. [72] using ^{1}H-MRS. Twelve subjects within the age range 60–80 years, supplemented for 12 weeks with $(2 \times 800$ mg$) \times 2 \cdot$day^{-1} CarnoSyn^{TM} SR tablets, showed a mean increase of 85% in M-Carn in m. gastrocnemius. Times-to-exhaustion in a constant-load submaximal test and an incremental test were positively correlated with the increase in M-Carn. It was concluded by the authors that maintaining a high M-Carn could be beneficial both in the immediate and longer term, especially if this encourages subjects to maintain a more active life.

As yet, few studies have been undertaken on the benefits of β-alanine in specific sports, or using exercise models relevant to specific sports, both in elite trained and untrained subjects. In Hill et al. [4], subjects supplemented with β-alanine for 4 and 10 weeks showed a 13 and 16.2% increase in total work done in a cycle capacity test performed at a constant load, mirroring increases of 58.8 and 80.1% in M-Carn in m. vastus lateralis. Confirmation of the results was subsequently provided by Sale et al. [71]. In van Thienen et al. [80], there was no effect of β-alanine on power output in a 10-min time trial undertaken immediately at the end of a simulated 110-min race. However, as noted, peak power output in a 30-s sprint after this was increased by 11% with β-alanine supplementation. In an examination of the changes in blood pH and oxygen uptake kinetics with 6 min of high-intensity cycling exercise, Baguet et al. [85] observed that exercise-induced acidosis was significantly reduced following β-alanine supplementation compared to placebo without affecting blood lactate. Further, that the time delay of the fast component (Td$_1$) of oxygen uptake kinetics was significantly reduced following β-alanine supplementation, although this did not lead to a reduction in oxygen deficit. The parameters of the slow component did not differ between groups. The authors concluded the increase in M-Carn with β-alanine supplementation was effective in attenuating the fall in blood pH during such exercise. Previously, Stout et al. [82] had observed an increase in the ventilatory threshold in female subjects performing a continuous incremental cycle test after 28 days of β-alanine, a result which equally could be explained by an attenuation of the fall in blood pH. However, the effect of β-alanine supplementation on ventilatory threshold was not borne out in a subsequent study by the same group, again in women and using an incremental test [86].

Also of interest to competitive sport, Baguet et al. [76] reported a strong positive correlation between 100, 500, 2,000 and 6,000 m rowing speed and the M-Carn content in 18 elite Belgium rowers. Subjects were retested over 2,000 m after 7 weeks of supplementation with either 5 g·day^{-1} β-alanine or placebo, and showed a positive correlation between the increase in rowing speed and the increase in M-Carn. This would seem to have possible applications both to rowing and kayak and canoe racing. More recently, Painelli et al. [87] have shown in highly trained Brazilian junior-standard swimmers a 1–2% improvement in 100 and 200 m freestyle swimming performance following supplementation with 3.2 g·day^{-1} for 1 week followed by 6.4 g·day^{-1} for 3 weeks of β-alanine supplied in the form of CarnoSyn^{TM} SR β-alanine tablets.

Conclusion

Based upon the results from chemical analysis of muscle biopsies (m. vastus lateralis) and ^{1}H-MRS measurements (m. gastrocnemius, m. soleus and m. tibialis anterior), it seems definite that supplementation of humans with β-alanine can result in a 50% or more increase in M-Carn within 4 weeks. Still higher increases, up to 80%, have been recorded when supplementation has been extended. When any single dose exceeds 10 mg·kg^{-1} body mass this may be associated with symptoms of paraesthesia, affecting the head and mainly the upper trunk, starting 15 min after β-alanine ingestion and subsiding after 1 h [48]. Symptoms of paraesthesia may be attenuated, enabling doses of ~20 mg·kg^{-1} body mass to be ingested without symptoms, if a sustained release (CarnoSyn^{TM} SR) formulation of β-alanine is used [59, 66, 72].

The maximum increase in M-Carn in humans achievable with β-alanine supplementation has yet to be established. However, forward extrapolation of the data from Hill et al. [4] and Kendrick et al. [12] suggests that this would occur in a longer timeframe than 10 weeks, even on a dose of 6.4 g·day^{-1}, whilst the percentage increase may be in excess of 100%.

It is definite that any increase in M-Carn will result in an increase in the intramuscular cell physicochemical buffering capacity over the exercise-induced pH transit range, by virtue of the pK$_a$ of the histidine-imidazole ring, unless this is compensated for by the loss of other physicochemical buffers. However, the only candidates here would be a molar equivalent loss in phosphate or histidine residues occurring in proteins (and thus a major loss in muscle protein). Neither of these seems at all likely. The quantitative importance of M-Carn to buffering remains to be elucidated but needs to take into account the change in buffering capacity with the onset of exercise due to the net release of phosphate in PCr. Measurements of muscle buffering capacity by acid or alkali titration of muscle homogenates are of little value here as they include contributions from extracellular proteins and involve the complete hydrolysis of PCr [88].

It remains to be established which areas of exercise are limited by pH decrease, but where this is the case it is likely that these will be benefitted by an increase in M-Carn. Such exercise regimens are also likely to benefit from oral sodium bicarbonate loading which offers a less direct, and possibly less effective means (certainly in the case of isometric exercise above 45% MVIC with partial to complete occlusion of the local blood supply) to attenuate muscle pH_i decrease. It follows that in this case the greater the increase in M-Carn, which at peak will also confer the greatest increase in muscle buffering capacity, then potentially the greater effect on exercise performance.

However, already it seems that maximal exercise of <60 s duration is little affected by M-Carn increase [75], with the possible exception when this is preceded by a further bout of exercise likely to affect pH_i [80]. Evidence is presented that exercise performance of a longer duration, but still of an intense nature (e.g., cycling, rowing, swimming, and repeated bouts of sprinting as in football and hockey), may be improved through β-alanine supplementation. Two further studies indicate improvements in exercise capacity in the elderly with β-alanine supplementation, which may be of particular importance to those wishing to engage in a more active lifestyle over a longer period of time [72, 83].

With the exception of vegetarians, all humans engage to some extent in dietary supplementation of normal hepatic β-alanine output through the consumption of carnosine and related HCDs in meat. Only recently has it become evident how important meat ingestion or direct β-alanine supplementation is to the M-Carn level in humans [7, 55]. As a result of the wide range in meat intake (from 0 g in vegans to perhaps 500 g·day^{-1} in high meat consumers), and the marked difference in the HCD content of different meats and the influence of cooking procedures [47, 56, 89, 90], the β-alanine content of current diets is estimated to vary from 0 to 1,000 mg·day^{-1}, with the upper value relevant to the traditional Mongolian diet [91] as an example of a nomadic people eating a diet composed almost mostly of meat (lamb, goat, chicken, horse and camel). As discussed by Wallimann et al. [22], in the context of creatine ingestion, fossil evidence indicates that diets for the major portion of hominid evolution in northern territories were predominantly composed of meat, from which we again estimate a possible dietary β-alanine intake of 1,000 mg/day or more. Even this estimate would be on the low side for a true carnivore with a body weight similar to that of a human where intakes of β-alanine may exceed 5 g/day. High levels of M-Carn would clearly have been advantageous in the course of human history, characterised as it has been by millennia of physical conflict and hard physical labour. Even a modest increase in M-Carn would likely have conferred advantage to an individual or even a tribe. In a similar way, elevation of M-Carn by β-alanine supplementation, especially against the background of lowered meat consumption currently favoured by society, may well enhance physical prowess in today's sporting warriors, as well benefitting those also engaged in more mundane but equally demanding tasks. Our current understanding is that competitive events involving high-intensity exercise of short to moderate duration will be amongst those likely to benefit from β-alanine supplementation. Table 1 provides examples of relevant events at the London 2012 Olympic Games and where the difference in performance time between gold, silver and bronze medal positions was, in most cases, less than 2%.

Table 1. Events at the 2012 London Olympics having exercise durations which suggest a potential benefit from supplementation with β-alanine

		Men			Women		
		1st (min:sec.00)	2nd (sec over 1st place)	3rd (sec over 2nd place)	1st (min:sec.00)	2nd (sec over 1st place)	3rd (sec over 2nd place)
Rowing	Pair	6:16.65	+4.46	+0.66	7:27.13	+2.73	+0.33
	Four	6:03.97	+1.22	+2.01	no event		
	Lightweight double four	6:02.84	+0.25	+0.07	no event		
	Eight	5:48.75	+1.23	+1.20	6:10.59	+1.47	+1.06
	Single sculls	6:57.82	+1.55	+3.91	7:54.37	+3.35	+0.32
	Double sculls	6:31.67	+1.13	+1.55	6:55.82	+2.73	+9.37
	Lightweight double sculls	6:37.17	+0.61	+3.08	7:09.30	+2.63	+0.16
	Quadruple sculls	5:42.48	+2.30	+0.44	6:35.93	+2.16	+2.54
Swimming	100 m breaststroke	0:58.46	+0.47	+1.03	1:05.47	+0.08	+0.99
	100 m backstroke	0:52.16	+0.76	+0.81	0:58.33	+0.35	+0.50
	100 m butterfly	0:51.21	+0.23	+0.23	0:55.98	+0.89	+0.96
	100 m freestyle	0:47.52	+0.01	+0.28	0:53.00	+0.38	+0.44
	200 m breaststroke	2:07.28	+0.15	+1.01	2:19.59	+1.13	+1.33
	200 m backstroke	1:53.41	+0.37	+0.53	2:04.06	+1.86	+2.49
	200 m butterfly	1:52.96	+0.05	+0.25	2:04.06	+1.19	+1.42
	200 m freestyle	1:43.14	+1.79	+1.79	1:53.61	+1.97	+2.20
	200 m individual medley	1:54.27	+0.63	+1.95	2:07.57	+0.58	+1.38
	400 m freestyle	3:40.14	+1.92	+4.55	4:01.45	+0.32	+1.56
	400 m individual medley	4:05.18	+3.68	+3.76	4:28.43	+2.84	+4.48

References

1 Bergstrom J, Furst P, Noree LO, Vinnars E: Intracellular free amino acid concentration in human muscle tissue. J Appl Physiol 1974;36:693–697.

2 Harris RC, Hultman E, Nordesjo LO: Glycogen, glycolytic intermediates and high energy phosphates in biopsy samples of musculus quadriceps femoris of man at rest. Methods and variance of values. Scand J Clin Lab Invest 1974;33:109-120.

3 Harris RC, Dunnett M, Greenhaff PL: Carnosine and taurine contents in individual fibres of human vastus lateralis muscle. J Sports Sci 1998;16:639–643.

4 Hill CA, Harris RC, Kim HJ, Harris BD, Sale C, Boobis LH, Kim CK, Wise JA: Influence of β-alanine supplementation on skeletal muscle carnosine concentrations and high-intensity cycling capacity. Amino Acids 2007;32:225–233.

5 Tallon MJ, Harris RC, Boobis LH, Fallowfield JL, Wise JA: The carnosine content of vastus lateralis is elevated in resistance trained bodybuilders. J Strength Cond Res 2005;19:725–729.

6 Kendrick IP, Kim HJ, Harris RC, Kim CK, Dang VH, Lam TQ, Bui TT, Wise JA: The effect of 4 weeks β-alanine supplementation and isokinetic training on carnosine concentrations in type I and II human skeletal muscle fibres. Eur J Appl Physiol 2009;106: 131–138.

7 Harris RC, Wise JA, Price KA, Kim HJ, Kim CK, Sale C: Determinants of muscle carnosine content. Amino Acids 2012;43:5–12.

8 Baguet A, Everaert I, Achten E, Thomis M, Derave W: The influence of sex, age and heritability on human skeletal muscle carnosine content. Amino Acids 2012;43:13–20.

9 Penafiel R, Ruzafa C, Monserrat F, Cremades A: Gender related differences in carnosine, anserine and lysine content of murine skeletal muscle. Amino Acids 2004;26:53–58.

10 Suzuki Y, Ito O, Takahashi H, Takamatsu K: The effect of sprint training on skeletal muscle carnosine in humans. Int J Sport Health Sci 2004;2:105–110.

11 Mannion AF, Jakeman PM, Willan PLT: Effects of isokinetic training of the knee extensors on high-intensity exercise performance and skeletal muscle buffering. Eur J Appl Physiol 1994;68:356–361.

12 Kendrick IP, Harris RC, Kim HJ, Kim CK, Dang VH, Lam TQ, Bui TT, Smith M, Wise JA: The effects of 10 weeks of resistance training combined with β-alanine supplementation on whole-body strength, force production, muscular endurance and body composition. Amino Acids 2008;34:547–554.

13 Kendrick IP: Effect of β-alanine supplementation and training on skeletal muscle carnosine synthesis. PhD thesis, University of Chichester, UK, 2011.

14 Parkhouse WS, McKenzie DC, Hochachka PW, Ovalle WK: Buffering capacity of deproteinized human vastus lateralis muscle. J Appl Physiol 1985; 58:14–17.

15 Hultman E, Sahlin K: Acid-base balance during exercise. Exerc Sport Sci Rev 1980;8:41–128.

16 Sahlin K, Harris RC, Nylind B, Hultman E: Lactate content and pH in muscle obtained after dynamic exercise. Pflugers Arch 1976;367:143–149.

17 Pan JW, Hamm JR, Hetherington HP, Rothman DL, Shulman RG: Correlation of lactate and pH in human skeletal muscle after exercise by ^{1}H-NMR. Magn Reson Med 1991;20:57–65.

18 Spriet LL, Lindinger MI, McKelvie RS, Heigenhauser GJF, Jones NL: Muscle glycogenolysis and H^+ concentration during maximal intermittent cycling. J Appl Physiol 1989;66:8–13.

19 Harris RC, Edwards RH, Hultman E, Nordesjo LO, Nylind B, Sahlin K: The time course of phosphorylcreatine resynthesis during recovery of the quadriceps muscle in man. Pflugers Arch 1976;367: 137–142.

20 Sahlin K: Metabolic factors in fatigue. Sports Med 1992;13:99–107.

21 Sahlin K, Harris RC: The creatine kinase reaction: a simple reaction with functional complexity. Amino Acids 2011;40:1363–1367.

22 Wallimann T, Tokarska-Schlattner M, Schlattner U: The creatine kinase system and pleiotropic effects of creatine. Amino Acids 2011;40:1271–1296.

23 Trivedi B, Daniforth WH: Effect of pH on the kinetics of frog muscle phosphofructokinase. J Biol Chem 1966;241:4110–4112.

24 Donaldson SKB, Hermansen L: Differential direct effects of H^+- and Ca^{2+}-activated force of skinned fibres from the soleus, cardiac, adductor magnus muscle of rabbits. Pflugers Arch 1978;376:55–65.

25 Fabiato A, Fabiato F: Effects of pH on the myofilaments and the sarcoplasmic reticulum of skinned cells from cardiac and skeletal muscles. J Physiol 1978;276:233–235.

26 Bate-Smith EC: The buffering of muscle in rigour: protein, phosphate and carnosine. J Physiol 1938; 92:336–343.

27 Harris RC, Marlin DJ, Dunnett M, Snow DH, Hultman E: Muscle buffering capacity and dipeptide content in the thoroughbred horse, greyhound dog and man. Comp Biochem Physiol 1990;97A: 249–251.

28 Tallon MJ, Harris RC, Maffulli N, Tarnopolsky MA: Carnosine, taurine and enzyme activities of human skeletal muscle fibres from elderly subjects with osteoarthritis and young moderately active subjects. Biogerontology 2007;8:129–137.

29 Hipkiss AR, Carmichael PL, Zimmermann B: Metabolism of crystalline fragments in cell-free extracts of bovine lens: effects of ageing and oxygen free-radicals. Acta Biol Hung 1993;42:243–263.

30 Hipkiss AR, Michaelis J, Syrris P: Non-enzymatic glycosylation of the dipeptide L-carnosine, a potential anti-protein-crosslinking agent. FEBS Lett 1995; 28:81–85.

31 Hipkiss AR, Brownson C, Carrier MJ: Carnosine, the anti-ageing, anti-oxidant dipeptide, may react with protein carbonyl groups. Mech Ageing Dev 2001;15:1431–1445.

32 Boldyrev AA, Dupin AM, Bunin AY, Babizhaev MA, Severin SE: The antioxidative properties of carnosine, a natural histidine-containing dipeptide. Biochem Int 1987;15:1105–1113.

33 Boldyrev AA: Does carnosine possess direct antioxidant activity? Int J Biochem 1993;25:1101–1107.

34 Lamont C, Miller DJ: Calcium sensitizing action of carnosine and other endogenous imidazoles in chemically skinned striated muscle. J Physiol 1992; 454:421–434.

35 Batrukova MA, Rubtsov AM: Histidine-containing dipeptides as endogenous regulators of the activity of sarcoplasmic reticulum Ca-release channels. Biochim Biophys Acta 1997;1324:142–150.

36 Rubtsov AM: Molecular mechanisms of regulation of the activity of sarcoplasmic reticulum Ca-release channels (ryanodine receptors), muscle fatigue, and Severin's phenomenon. Biochemistry 2001;66: 1132–1143.

37 Dutka TL, Lamb GD: Effect of carnosine on excitation-contraction coupling in mechanically-skinned rat skeletal muscle. J Muscle Res Cell Motil 2004;25: 203–213.

38 Dutka TL, Lamboley CR, McKenna MJ, Murphy RM, Lamb GD: Effects of carnosine on contractile apparatus Ca^{2+} sensitivity and sarcoplasmic reticulum Ca^{2+} release in human skeletal muscle fibers. J Appl Physiol 2012;112:728–736.

39 Hultman E, Del Canale S, Sjöholm H: Effect of induced metabolic acidosis on intracellular pH, buffer capacity and contraction force of human skeletal muscle. Clin Sci (Lond) 1985;69:505–510.

40 Ng RH, Marshall FD: Regional and subcellular distribution of homocarnosine-carnosine synthetase in the central nervous system of rats. J Neurochem 1978;30:187–190.

41 Skaper SD, Das S, Marshall FD: Some properties of a homocarnosine-carnosine synthetase isolated from rat brain. J Neurochem 1973;21:1429–1445.

42 Horinishi H, Grillo M, Margolis FL: Purification and characterization of carnosine synthetase from mouse olfactory bulbs. J Neurochem 1978;31: 909–919.

43 Fritzson P: The catabolism of ^{14}C-labeled uracil, dihydrouracil, and β-ureidopropionic acid in rat liver slices. J Biol Chem 1957;226:223–228.

44 Matthews MM, Traut TW: Regulation of N-carbamoyl-β-alanine amidohydrolase, the terminal enzyme in pyrimidine catabolism, by ligand-induced change in polymerization. J Biol Chem 1987;262:7232–7237.

45 Hama T, Tamaki N, Miyamoto F, Kita M, Tsunemori F: Intestinal absorption of β-alanine, anserine and carnosine in rats. J Nutr Sci Vitaminol (Tokyo) 1976;22:147–157.

46 Gardner ML, Illingworth KM, Kelleher J, Wood D: Intestinal absorption of the intact peptide carnosine in man, and comparison with intestinal permeability to lactulose. J Physiol 1991;439:411–422.

47 Park YJ, Volpe SL, Decker EA: Quantitation of carnosine in human plasma after dietary consumption of beef. J Agric Food Chem 2005;53:4736–4739.

48 Harris RC, Tallon MJ, Dunnett M, Boobis LH, Coakley J, Kim HJ, Fallowfield JL, Chester CA, Sale C, Wise JA: The absorption of orally supplied ß-alanine and its effect on muscle carnosine synthesis in human vastus lateralis. Amino Acids 2006;30: 279–289.

49 Asatoor AM, Bardon JK, Lant AF, Milne MD, Navab F: Intestinal absorption of carnosine and its constituent amino acids in man. Gut 1970;11:250–254.

50 Jackson MC, Kucera CM, Lenney JF: Purification and properties of human serum carnosinase. Clin Chim Acta 1991;196:193–205.

51 Sadikali F, Darwish R, Watson WC: Carnosinase activity of human gastrointestinal mucosa. Gut 1975;16:585–589.

52 Miyamoto Y, Nakamura H, Hoshi T, Ganapathy V, Leibach FH: Uphill transport of β-alanine in intestinal brush-border membrane vesicles. Am J Physiol 1990;259:G372–G379.

53 Ramamoorthy S, Leibach FH, Mahesh VB, Han H, Yang-Feng T, Blakely RD, Ganapathy V: Functional characterization and chromosomal localization of a cloned taurine transporter from human placenta. Biochem J 1994;300:893–900.

54 Bakardjiev A, Bauer K: Transport of β-alanine and biosynthesis of carnosine by skeletal muscle cells in primary culture. Eur J Biochem 1994;225:617–623.

55 Harris RC, Jones G, Hill CH, Kendrick IP, Boobis L, Kim CK, Kim HJ, Dang VH, Edge J, Wise JA: The carnosine content of v. lateralis in vegetarians and omnivores. FASEB J 2007;21:769.20.

56 Jones G: Imidazole dipeptides: dietary sources and factors affecting uptake and muscle content. PhD thesis, University of Chichester, UK, 2011.

57 Everaert I, Mooyaart A, Baguet A, Zutinic A, Baelde H, Achten E, Taes Y, De Heer E, Derave W: Vegetarianism, female gender and increasing age, but not CNDP1 genotype, are associated with reduced muscle carnosine levels in humans. Amino Acids 2011;40:1221–1229.

58 Harris RC, Jones G, Wise JA: The plasma concentration-time profile of β-alanine using a controlled release formulation (CarnoSyn™). FASEB J 2008;22:701.9.

59 Stellingwerff T, Anwander H, Egger A, Buehler T, Kreis R, Decombaz J, Boesch C: Effect of two β-alanine dosing protocols on muscle carnosine synthesis and washout. Amino Acids 2012;42: 2461–2472.

60 Baguet A, Reyngoudt H, Pottier A, Everaert I, Callens S, Achten E, Derave W: Carnosine loading and washout in human skeletal muscles. J Appl Physiol 2009;106:837–842.

61 Stellingwerff T, Decombaz J, Harris RC, Boesch C: Optimizing human in vivo dosing and delivery of β-alanine supplements for muscle carnosine synthesis. Amino Acids 2012;43:57–65.

62 Hipkiss AR: Carnosine and protein carbonyl groups: a possible relationship. Biochemistry (Moscow) 2000;65:771–778.

63 Guiotto A, Calderan A, Ruzza P, Borin G: Carnosine and carnosine-related antioxidants: a review. Curr Med Chem 2005;12:2293–2315.

64 Harris RC, Dunnett M, Snow DH: Muscle carnosine content is unchanged during maximal intermittent exercise. Equine Exerc Physiol 1991;3:257–261.

65 Dunnett M, Harris RC, Dunnett CE, Harris PA: Plasma carnosine concentration: diurnal variation and effects of age, exercise and muscle damage. Equine Vet J Suppl 2002;34:283–287.

66 Decombaz J, Beaumont M, Vuichoud J, Bouisset F, Stellingwerff T: Effect of slow-release β-alanine tablets on absorption kinetics and paresthesia. Amino Acids 2012;43:67–76.

67 Mori M, Gahwiler BH, Gerber U: Beta-alanine and taurine as endogenous agonists at glycine receptors in rat hippocampus in vitro. J Physiol 2002;15:191–200.

68 Tokutomi N, Kaneda M, Akaike N: What confers specificity on glycine for its receptor site? Br J Pharmacol 1989;97:353–360.

69 Wang DS, Zhu HL, Li JS: Beta-alanine acts on glycine receptors in the rat sacral dorsal commissural neurons. Int J Neurosci 2003;113:293–305.

70 Crozier RA, Ajit SK, Kaftan EJ, Pausch MH: MrgD activation inhibits KCNQ/M-currents and contributes to enhanced neuronal excitability. J Neurosci 2007;27:4492–4496.

71 Sale C, Saunders B, Hudson S, Wise JA, Harris RC, and Sunderland CD: Effect of β-alanine plus sodium bicarbonate on high-intensity cycling capacity. Med Sci Sports Exerc 2011;43:1972–1978.

72 Del Favero S, Roschel H, Solis MY, Hayashi AP, Artioli GG, Otaduy MC, Benatti FB, Harris RC, Wise JA, Leite CC, Pereira RM, de Sá-Pinto AL, Lancha-Junior AH, Gualano B: Beta-alanine (Carnosyn™) supplementation in elderly subjects (60–80 years): effects on muscle carnosine content and physical capacity. Amino Acids 2012;43:49–56.

73 Saunders B, Sale C, Harris RC, Sunderland C: Effect of β-alanine supplementation on repeated sprint performance during the Loughborough Intermittent Shuttle Test. Amino Acids 2012;43:39–47.

74 Saunders B, Sunderland C, Harris RC, Sale C: Beta-alanine supplementation improves YoYo intermittent recovery test performance. J Int Soc Sports Nutr 2012;9:39.

75 Hobson RM, Saunders B, Ball G, Harris RC, Sale C: Effects of β-alanine supplementation on exercise performance: a review by meta-analysis. Amino Acids 2012;43:25–37.

76 Baguet A, Bourgois J, Vanhee L, Achten E, Derave W: Important role of muscle carnosine in rowing performance. J Appl Physiol 2010;109:1096–1101.

77 Sale C, Hill CA, Ponte J, Harris RC: Beta-alanine supplementation improves isometric endurance of the knee extensor muscles. J Intl Soc Sports Nutr 2012;9:26.

78 Ahlborg B, Bergström J, Ekelund L, Guarnieri G, Harris RC, Hultman E, Nordesjö L: Muscle metabolism during isometric exercise performed at constant force. J Appl Physiol 1972;33:224–228.

79 Jagim AR, Wright GA, Brice AG, Doberstein ST: Effects of β-alanine supplementation on sprint endurance. J Strength Cond Res 2012 (E-pub ahead of print).

80 Van Thienen R, van Proeyen K, van den Eynde B, Puype J, Lefere T, Hespel P: Beta-alanine improves sprint performance in endurance cycling. Med Sci Sports Exerc 2009;41:898–903.

81 Stout JR, Cramer JT, Mielke M, O'Kroy J, Torok DJ, Zoeller RF: Effects of twenty-eight days of β-alanine and creatine monohydrate supplementation on the physical working capacity at neuromuscular fatigue threshold. J Strength Cond Res 2006;20:928–931.

82 Stout JR, Cramer JT, Zoeller RF, Torok D, Costa P, Hoffman JR, Harris RC, O'Kroy J: Effects of β-alanine supplementation on the onset of neuromuscular fatigue and ventilatory threshold in women. Amino Acids 2007;32:381–386.

83 Stout JR, Graves BS, Smith AE, Hartman MJ, Cramer JT, Beck TW, Harris RC: The effect of β-alanine supplementation on neuromuscular fatigue in elderly (55–92 years): a double-blind randomized study. J Int Soc Sports Nutr 2008;5:21–26.

84 Zoeller RF, Stout JR, O'Kroy J, Torok D, Mielke M: Effects of 28 days of β-alanine and creatine monohydrate supplementation on aerobic power, ventilatory and lactate thresholds and time to exhaustion. Amino Acids 2007;33:505–510.

85 Baguet A, Koppo K, Pottier A, Derave W: Beta-alanine supplementation reduces acidosis but not oxygen uptake response during high-intensity cycling exercise. Eur J Appl Physiol 2010;108:495–503.

86 Smith AE, Stout JR, Kendall KL, Fukuda DH, Cramer JT: Exercise-induced oxidative stress: the effects of β-alanine supplementation in women. Amino Acids 2012;43:1197–1205.

87 Painelli V, Roschel H, de Jesus F, Sale C, Harris RC, Galves VF, de Oliveira N, do Carmo CA, Solis MY, Benatti FB, Gualano B, Lancha AH, Artioli GG: Effects of β-Alanine combined with sodium bicarbonate on swimming performance (submitted, 2012).

88 Marlin DJ, Harris RC: Titrimetric determination of muscle buffering capacity (β-mtitr) in biopsy samples. Equine Vet J 1991;23:193-197.

89 Purchas RW, Rutherfurd SM, Pearce PD, Vather R, Wilkinson BH: Concentrations in beef and lamb of taurine, carnosine, coenzyme Q_{10}, and creatine. Meat Sci 2004;66:629–637.

90 Yeum KJ, Orioli M, Regazzoni L, Carini M, Rasmussen H, Russell RM, Aldini G: Profiling histidine dipeptides in plasma and urine after ingesting beef, chicken or chicken broth in humans. Amino Acids 2010;38:847–858.

91 Cavendish M: World and Its Peoples: Eastern and Southern Asia. Tarrytown/NY, Marshall Cavendish Corp, 2007, pp 268–269.

Roger C. Harris
Junipa Ltd
Newmarket, Suffolk CB8 8HD (UK)
Tel. +44 79 13 79 89 48
E-Mail junipa@ymail.com

Lamprecht M (ed): Acute Topics in Sport Nutrition.
Med Sport Sci. Basel, Karger, 2013, vol 59, pp 18–28

Arginine and Citrulline Supplementation in Sports and Exercise: Ergogenic Nutrients?

Antoni Sureda · Antoni Pons

Laboratory of Physical Activity Science, Research Group on Community Nutrition and Oxidative Stress, Department of Biologia Fonamental i Ciències de la Salut, University of the Balearic Islands, Palma de Mallorca, Illes Balears, Spain

Abstract

Dietary L-citrulline malate supplements may increase levels of nitric oxide (NO) metabolites, although this response has not been related to an improvement in athletic performance. NO plays an important role in many functions in the body regulating vasodilatation, blood flow, mitochondrial respiration and platelet function. L-Arginine is the main precursor of NO via nitric oxide synthase (NOS) activity. Additionally, L-citrulline has been indicated to be a second NO donor in the NOS-dependent pathway, since it can be converted to L-arginine. The importance of L-citrulline as an ergogenic support derives from the fact that L-citrulline is not subject to pre-systemic elimination and, consequently, could be a more efficient way to elevate extracellular levels of L-arginine by itself. L-Citrulline malate can develop beneficial effects on the elimination of NH_3 in the course of recovery from exhaustive muscular exercise and also as an effective precursor of L-arginine and creatine. Dietary supplementation with L-citrulline alone does not improve exercise performance. The ergogenic response of L-citrulline or L-arginine supplements depends on the training status of the subjects. Studies involving untrained or moderately healthy subjects showed that NO donors could improve tolerance to aerobic and anaerobic exercise. However, when highly-trained subjects were supplemented, no positive effect on performance was indicated.

Copyright © 2012 S. Karger AG, Basel

L-Citrulline is a non-essential amino acid which participates in central reactions in the urea cycle. Its name is derived from *citrullus*, the Latin word for watermelon, from which it was first isolated in 1930. Citrulline can act as a precursor for the synthesis of arginine which is the substrate for nitric oxide synthases (NOS). NOS enzymes catalyze a complex enzymatic reaction leading to NO formation from L-arginine and molecular oxygen [1]. Nitric oxide (NO) is a labile lipid soluble gas that regulates important functions as a mediator in noradrenergic and non-cholinergic neurotransmission, in learning and memory, synaptic plasticity, and neuroprotection [2]. Three isoforms of NOS exist: neuronal (nNOS, NOS 1), inducible (iNOS, NOS 2) and endothelial (eNOS, NOS 3), which convert L-arginine to NO [1]. An alternative NOS-independent pathway of NO synthesis was discovered based on the simple reduction of nitrate and nitrite [3, 4], the main oxidation products of NO.

In sport physiology, NO has also received much interest, and supplements of NO are thought to be an ergogenic aid [5]. This fact is based on the evidence that NO is an important modulator of blood flow and mitochondrial respiration during physical exercise [6]. In addition, it is suggested that the increase in blood flow derived from NO synthesis may improve recovery processes of the activated tissues [7]. Some of these studies showed that dietary NO supplements may enhance human performance in healthy subjects [8–10], but other investigations did not find any positive effect [11–13]. One reason to explain this fact could be the great methodological differences between studies. Duration of treatment, exercise protocol and training status differ significantly between studies, making a comparison between them difficult. Additionally, many studies have used NO donors in combination with other components such as malate, glutamate, aspartate, etc. in an attempt to increase the bioavailability of NO donors. This fact makes it more difficult because some of these additional products may participate in the independent NO synthesis pathways in the body.

Accordingly, this chapter focuses on citrulline and arginine as donors of NO synthesis and elucidates the effect of citrulline/arginine supplements on human performance. Scientific articles were retrieved based on an extensive search in MEDLINE (1980–2011) and Google Scholar (1990–2011) databases. Computer search engines used the following combined keywords: 'L-citrulline', 'malate', 'L-arginine', 'nitrate', 'supplementation', 'nitric oxide', 'exercise' and 'performance'. After using these initial keywords, the search engines were limited to human studies excluding research with animals and also humans in pathological states. Only articles related to the effects of dietary ingredients linked with NO and performance in response to exercise were considered.

L-Citrulline: Sources and Metabolism

In addition to the dietary source, L-citrulline is also produced endogenously by two main pathways: (1) synthesized from glutamine that produces ornithine in enterocytes, and then by condensation of ornithine and carbamoylphosphate in a reaction catalyzed by ornithine carbamoyltransferase [14, 15]; (2) produced by the conversion of L-arginine to NO in a reaction catalyzed by NOS enzymes. The normal value of L-citrulline reported in healthy populations is approximately 25 μmol$\cdot$l^{-1} [16], although lower values have recently been found (10–15 μmol$\cdot$l^{-1}) in professional cyclists [17].

Dietary interest for this amino acid has substantially increased during the last decade due to the importance of L-citrulline as a precursor of L-arginine [18, 19].

This is interesting because L-arginine is subject to extensive pre-systemic and systemic elimination by arginase in the gut wall and liver [20, 21], whereas L-citrulline is not subject to pre-systemic elimination. For this reason, it has been indicated that systemic administration of L-citrulline could be a more efficient way to elevate extracellular levels of L-arginine by itself [22]. Dietary L-citrulline is taken up and released by enterocytes in the portal circulation, bypasses the metabolism by periportal hepatocytes and is transported to the kidneys where around 80% is catabolized to L-arginine by proximal tubule [23]. Apart from the function as a precursor of L-arginine, it is known that L-citrulline is an essential component participating in the urea cycle in the liver [24]. The ergogenic effects of L-citrulline could be directly related to its capacity to increase L-arginine levels and consequently increase the substrate for NO and creatine synthesis.

Ergogenic Effect of L-Citrulline Supplements
Most studies carried out with L-citrulline were performed in combination with malate. Only one study was carried out involving L-citrulline supplementation alone without addition of other products. In this study, Hickner et al. [25] evaluated the effects of an incremental treadmill test to exhaustion following either placebo or citrulline ingestion (3 g 3 h before test, or 9 g over 24 h prior to testing) in young healthy subjects. Contrary to the hypothesis of the authors, the results showed that L-citrulline supplementation impaired exercise by a reduction in treadmill time following L-citrulline ingestion. To explain this unexpected response it was suggested that L-citrulline ingestion may reduce NO-mediated pancreatic insulin secretion or increase insulin clearance.

The first study to assess the effect of L-citrulline in combination with malate analyzed the effects of L-citrulline malate (6 g·day^{-1} × 12 days) on the accumulation of ammonia during a maximum exercise test [26]. Serum ammonia levels at the end of the test were higher than baseline values but there were no significant differences between the placebo and treatment groups. However, serum ammonia levels were significantly lower in the L-citrulline group during the recovery period. This study disclosed no difference between the two groups in lactate levels.

Two studies performed in our group showed an increase in plasma NO metabolites – such as nitrite, creatinine, ornithine and urea – in endurance-trained subjects after a cycling competition, returning to basal values after 3 h of recovery. These cyclists were supplemented with only one dose of L-citrulline malate (6 g) 2 h before exercise [17, 27]. In addition, an increase in plasma arginine availability was found (table 1) which was linked with substrate for NO synthesis, as well as an increase in the ability of polymorphonuclear neutrophils to produce ROS after activation with opsonized zymosan [17]. Polymorphonuclear neutrophils play an important role in the defense against infections, in the inflammatory response, and muscle repair and regeneration [28, 29]. L-Citrulline malate supplementation can also enhance the use of amino

Table 1. Effects of acute L-citrulline malate supplementation; adapted from Sureda et al. [9, 10]

	Basal	After race	3-Hour recovery
L-Citrulline, mol/l			
Control	22.4±3.1	25.2±5.7	14.2±2.0
Citrulline	24.2±3.7	66.2±22.1*,#	23.7±10.4
L-Arginine, mol/l			
Control	43.2±6.0	34.7±5.5	27.0±4.3
Citrulline	48.7±8.5	109±38*,#	40.7±13.0
L-Ornithine, mol/l			
Control	53.6±5.9	57.5±1.3	86.0±34.7
Citrulline	55.5±8.2	140±41*,#	88.5±13.5
Urea, mg/dl			
Control	35.8±2.0	50.2±3.7*	46.8±3.6*
Citrulline	41.0±2.6	65.5±3.9*,#	59.5±3.6*,#
Nitrite, nmol/ml			
Control	1.76±0.23	2.10±0.34	1.98±0.21
Citrulline	1.73±0.24	2.32±0.31*	2,23±0.28

Effects of exercise and L-citrulline malate supplementation (6 g, 2 h prior exercise) on plasma urea, amino acids of the urea cycle and nitrite determined before a 137-km cycling stage in basal conditions, immediately after the race, and after 3 h of recovery. Two-way ANOVA.
* Significant differences with respect to basal values.
# Significant differences between control and citrulline-supplemented groups. Control n = 9, citrulline supplemented n = 8. Values are mean ± SEM, $p < 0.05$.

acids, especially branched-chain amino acids during exercise. Unfortunately, it was not possible to associate these findings with variables of exercise performance due to the characteristics of the study design. Many factors such as strategy, environmental conditions, nutrition, drafting and breakdown of material can affect the results during sport field events, thus limiting the use of these data to evaluate the association between dietary supplement and performance.

Another recent study by Pérez-Guisado et al. [30] showed that a single dose of L-citrulline and malate (8 g) increased athletic performance in terms of work capacity, in high-intensity anaerobic exercises with short rest times, by an average of 19% measured with pectoral training session protocols. However, this finding cannot be related to NO delivery as plasma NO markers were not determined in this study [30]. However, about 15% of subjects had a feeling of stomach discomfort.

In another group supplemented with L-citrulline malate (6 g·day^{-1} × 16 days), the high-energy phosphate metabolism was analyzed by ^{31}P magnetic resonance spectroscopy during a finger flexion exercise. The study concluded that supplementation resulted in a significant increase (34%) in the rate of oxidative ATP production during exercise, and a 20% increase in the rate of phosphocreatine recovery after exercise [31]. However, there is a great limitation to this research because there was no placebo group or blind condition in the design. The improved aerobic function detected may have been the result of an enhanced malate supply activating ATP production from the tricarboxylic acid (TCA) cycle through anaplerotic reactions, as has been suggested [31]. L-Malate plays a central role in the transfer from cytosolic nicotinamide adenine dinucleotide (NADH) to mitochondrial NADH, a possible proton exchange [32]. L-Malate could increase the efficiency of the NADH shuttle and energy synthesis by enhancing the activities of enzymes related to the malate/ aspartate shuttle [33]. A study was carried out in humans supplemented only with malate. However, although there was up to a fourfold increase in the concentration of the TCA cycle intermediates at the start of exercise, the concentration of these intermediates did not limit the rate of TCA cycle flux or, thereby, the oxidative metabolism [34].

Taking all this overview together, it is evident that there is a lack of data linking an increase in exercise performance with an increase in NO production derived from L-citrulline supplementation. During intense exercise, there is an augmented production of ammonia, a product that accumulates in skeletal muscle when AMP is deaminated to inosine monophosphate during the resynthesis of ATP, which is thought to be one of the causes of exercise-induced fatigue. L-Citrulline supplementation could facilitate ammonia detoxification during exercise and improve the recovery process. Performance enhancement reported by L-citrulline in combination with malate could be explained by the interaction of these molecules in other metabolic pathways that are independent of NO production. For instance, L-citrulline increases levels of plasma L-arginine, and could also indirectly enhance the synthesis of creatine, since L-arginine supplementation has been reported to stimulate an rise in intramuscular creatine concentration [35]. Therefore, this mechanism may improve the response to anaerobic exercise. In addition, malate may be involved in the beneficial effects on energy production because it is an intermediate in the TCA cycle [25, 36]. It has been suggested that hyperactivation of aerobic ATP production coupled to a reduction in anaerobic energy supply may contribute to the reduction in fatigue sensation reported by the subjects [37].

In summary, the conclusions that we can extract according to the studies performed with L-citrulline as dietary supplementation in sport are: (1) Dietary supplementation with L-citrulline alone does not improve exercise performance. (2) Dietary L-citrulline malate supplements may increase levels of NO metabolites, although this response has not been directly related to an improvement in athletic performance. (3) L-Citrulline malate may develop beneficial effects on the elimination of NH_3 in the course of recovery from exhaustive muscular exercise and also as an effective precursor of L-arginine and creatine.

L-Arginine: Sources and Metabolism. Synthesis of NO from NOS-Dependent Pathway

L-Arginine is a basic amino acid – considered a conditional essential proteinogenic amino acid – that is a natural constituent of dietary proteins. Free L-arginine within the body is derived from the diet, endogenous synthesis and turnover of proteins. Its content is relatively high in seafood, watermelon juice, nuts, seeds, algae, meats, rice protein concentrate and soy protein isolate [38]. The typical dietary intake of L-arginine is approximately of 4–5 g/day. However, substantial amounts of orally administered arginine do not enter the systemic circulation in adults because about 40% of dietary L-arginine is degraded by the small intestine in the first-pass metabolism [33, 39]. L-Arginine could be endogenously synthesized, mainly in the kidney, where L-arginine is formed from L-citrulline [40]. The liver is also able to synthesize considerable amounts of L-arginine, although this is completely reused in the urea cycle [40]. Normal plasma L-arginine concentrations depend upon the age of the individual, and its homeostasis is primarily achieved via its catabolism [39]. Plasma levels of arginine in healthy adults are between 80 and 120 μmol/l. In healthy adults the level of endogenous synthesis is sufficient to cover metabolic demands. However, in cases of catabolic stress, such as inflammation and infection or conditions involving dysfunction of the kidneys or small intestine, levels of endogenous synthesis may not suffice to meet metabolic demands.

L-Arginine participates in multiple metabolic fates. It is metabolically interconvertible with the amino acids proline and glutamate, but it also serves as a precursor for the synthesis of protein, NO, creatine, polyamines, agmatine, and urea [41]. L-Arginine amino acid takes part in the NOS-dependent pathway in a reaction catalyzed by specific NOS enzymes [2]. L-Arginine also participates in other metabolic pathways independent of NO synthesis. For instance, L-arginine is essential for the normal function of the urea cycle, in which ammonia is detoxified through its metabolism into urea [42]. L-Arginine is also a potent hormone secretagogue. L-Arginine infusion at rest increases plasma insulin, glucagon, growth hormone, prolactin and catecholamine concentrations [43]. Arginine serves as a precursor for creatine synthesis by providing guanidino groups. It combines with glycine, a reaction catalyzed by arginine:glycine aminotransferase, to form guanidinoacetate, which is then converted to creatine by guanidinoacetate methyltransferase.

Ergogenic Effect of L-Arginine Supplements
Several studies have analyzed the effect of L-arginine supplementation alone [11, 20, 44–48]. Two of these studies were carried out in young healthy but not well-trained subjects [45, 20]. In the first study, Koppo et al. [45] showed a significant increase of speed in phase II of pulmonary Vo_2 at the onset of moderate intensity endurance cycle exercise after 14 days of L-arginine supplementation (7.2 g × day^{-1}), enhancing the tolerance to endurance exercise. However, these findings were not linked with

NO synthesis since the above studies did not report data related to NO markers. Olek et al. [20] assessed the effect of an acute low dose of L-arginine (2 g) 60 min before exercise. The results reported no changes in total work performed or mean power output during Wingate cycle tests (30 s), or oxygen consumption and plasma levels of nitrate/nitrite [20] either. A third study performed in postmenopausal women [44] supplemented with higher doses of L-arginine (14.2 g $\times$ day^{-1}) for 6 months reported a significant increase in maximal power in relation with body mass (power $\times$ kg^{-1}) measured as peak jump power (counter-movement jump) [44].

Focusing our attention on well-trained athletes, four studies have assessed the effect of L-arginine supplementation in different types of athletic populations such as judo athletes [46, 48], tennis players [11] and cyclists [47]. Despite analyzing supplements for different durations (between 1 and 28 days) and doses (between 6 and 12 g), no benefit was indicated in parameters linked with performance. Moreover, the levels of some exercise metabolites (lactate and ammonia) remained unchanged after L-arginine supplementation compared with placebo [46, 48]. Analysis of plasma nitrate/nitrite as NO markers showed that they did not rise after dietary L-arginine ingestion [11, 46, 48].

Apart from dietary supplementation, other investigations have analyzed the effect of intravenous infusion of L-arginine in an attempt to increase its bioavailability [49, 50]. However, even though bioavailability of intravenous infusion of L-arginine could be high compared with dietary consumption, no positive effect was reported on performance parameters [49, 50].

It has also been reported that large amounts of arginine (2.3 g/day in a 70-kg man) are used for the production of creatine via the interorgan cooperation of kidneys, pancreas, liver, and skeletal muscle [51]. Much evidence shows that creatine has an antioxidative function [52], reduces inflammatory responses [53] and improves glucose tolerance [13] in humans. Although studies with creatine supplementation have evidenced increased muscle mass, strength, and power [54–56], there is a lack of studies correlating L-arginine supplements with creatine-enhanced athletic performance.

Several studies have found an improvement in exercise performance after L-arginine supplementation in combination with other components in untrained or moderately-trained subjects. Bailey et al. [9] showed that L-arginine (6 g $\times$ 3 days) in combination with other amino acids and vitamins induced a decrease in oxygen consumption in low-moderate bouts of exercise, and an increased time to exhaustion during an incremental cycling test [21]. Similarly, Camic et al. [10] found a rise in power output during an incremental test to exhaustion with a cycle ergometer when L-arginine (3 g) was administered in combination with grapeseed extract for 28 days. These surprising findings were related to an increase in the gas exchange threshold induced by dietary L-arginine supplementation [57, 58]. It was suggested that the attenuation of metabolic products such as potassium, ammonia and lactate may be the result of increased clearance from circulation related to NO synthesis and increased blood flow

[57]. However, another study concluded that an oral bolus (10 g) of arginine plus 10 g essential amino acids does not increase NO synthesis or muscle blood flow; neither does arginine enhance muscle protein synthesis either at rest or after resistance exercise [59]. Other studies have also reported benefits of a mixture of L-arginine supplements on strength and power performance in moderately-trained subjects. Campbell et al. [60] indicated a significant increase in one repetition maximum (1-RM) bench press, as well as peak power during a 30-second Wingate test after L-arginine supplementation (6 g·day^{-1} × 56 days) in combination with α-ketoglutarate. Furthermore, Buford et al. [61] and Stevens et al. [62] showed that an acute dose of L-arginine (6 g of L-arginine) in the form of α-ketoisocaproic increased the mean power performed during Wingate tests (10 s) and work sustained during continuous isokinetic concentric/eccentric knee extension repetitions, respectively.

In well-trained athletes, two studies have assessed dietary L-arginine supplementation in combination with aspartate. In the first one, Colombani et al. [63] supplemented (15 g·day^{-1} × 14 days) endurance-trained runners. They showed that the plasma level of somatotropic hormone,, glucagon, urea, and arginine were significantly increased, while the levels of plasma amino acids were significantly reduced following a marathon run. The conclusion of this study was that there was no metabolic or performance benefit derived from L-arginine. Similar findings were reported by Abel et al. [64] who supplemented endurance-trained cyclists with L-arginine and aspartate at high (5.7 g of L-arginine; 8.7 g of aspartate) and low (2.8 g of L-arginine; 2.2 g of aspartate) doses for 28 days. After an incremental endurance exercise test, no modification was found in endurance performance or in metabolic parameters [64]. Therefore, including all studies with L-arginine supplementation alone and with other components, no study in well-trained athletes reported benefits on performance [10, 19, 39, 50, 60, 63]. These results in well-trained athletes could be explained by the physiology and metabolic adaptation derived from chronic physical training. The effect of exercise training on the enhancement of endothelial function has been well established [65]. Repetitive exercise over weeks results in an upregulation of eNOS activity [66]. Perhaps benefits in pulmonary, cardiovascular and neuromuscular systems induced by long-term training may overcome any potential effects of dietary L-arginine supplementation in well-trained athletes.

In addition, in most studies there is a lack of data concerning NO metabolites. Consequently, at present, it is not possible to relate enhancement in exercise performance derived from L-arginine supplementation with NO synthesis. Some benefits shown in the above studies could be related to other metabolic pathways independent of NO synthesis, as well as the other ingredients included in L-arginine supplements. For instance, there is evidence that L-arginine supplementation in combination with glutamate and aspartate is effective at reducing blood levels of ammonia [67, 68] and also blood lactate [50] during exercise.

In summary, current evidence concerning L-arginine supplementation in sport performance suggests that: (1) L-arginine, mainly in combination with other components, could induce some benefit in untrained or moderately-trained subjects, improving tolerance to aerobic and anaerobic physical exercise. However, as the studies were not well defined, a relationship between dietary L-arginine supplementation and NO synthesis and the benefit in exercise performance shown in some studies could be derived from other supplement ingredients, as well as other NO-independent metabolic pathways which L-arginine participates in. (2) In well-trained athletes there is a lack of data indicating that L-arginine supplementation induces benefits in performance.

References

1 Schmidt HH, Nau H, Wittfoht W, Gerlach J, Prescher KE, Klein MM, Niroomand F, Bohme E: Arginine is a physiological precursor of endothelium-derived nitric oxide. Eur J Pharmacol 1988; 154:213–216.

2 Moncada S, Higgs A: The L-arginine-nitric oxide pathway. N Engl J Med 1993;329:2002–2012.

3 Benjamin N, O'Driscoll F, Dougall H, Duncan C, Smith L, Golden M, McKenzie H: Stomach NO synthesis. Nature 1994;368:502.

4 Lundberg JO, Weitzberg E, Lundberg JM, Alving K: Intragastric nitric oxide production in humans: measurements in expelled air. Gut 1994;35:1543–1546.

5 Petroczi A, Naughton DP: Potentially fatal new trend in performance enhancement: a cautionary note on nitrite. J Int Soc Sports Nutr 2010;7:25.

6 Shen W, Xu X, Ochoa M, Zhao G, Wolin MS, Hintze TH: Role of nitric oxide in the regulation of oxygen consumption in conscious dogs. Circ Res 1994;75: 1086–1095.

7 Bloomer RJ: Nitric oxide supplements for sports. J Strength Cond Res 2010;32:14–20.

8 Bailey SJ, Winyard P, Vanhatalo A, Blackwell JR, Dimenna FJ, Wilkerson DP, Tarr J, Benjamin N, Jones AM: Dietary nitrate supplementation reduces the O_2 cost of low-intensity exercise and enhances tolerance to high-intensity exercise in humans. J Appl Physiol 2009;107:1144–1155.

9 Bailey SJ, Winyard PG, Vanhatalo A, Blackwell JR, DiMenna FJ, Wilkerson DP, Jones AM: Acute L-arginine supplementation reduces the O_2 cost of moderate-intensity exercise and enhances high-intensity exercise tolerance. J Appl Physiol 2010;109: 1394–1403.

10 Camic CL, Housh TJ, Zuniga JM, Hendrix RC, Mielke M, Johnson GO, Schmidt RJ: Effects of arginine-based supplements on the physical working capacity at the fatigue threshold. J Strength Cond Res 2010;24:1306–1312.

11 Bescos R, Gonzalez-Haro C, Pujol P, Drobnic F, Alonso E, Santolaria ML, Ruiz O, Esteve M, Galilea P: Effects of dietary L-arginine intake on cardiorespiratory and metabolic adaptation in athletes. Int J Sport Nutr Exerc Metab 2009;19:355–365.

12 Bescos R, Rodriguez FA, Iglesias X, Ferrer MD, Iborra E, Pons A: Acute administration of inorganic nitrate reduces VO_2 peak in endurance athletes. Med Sci Sports Exerc 2011;43:1979–1986.

13 Gualano B, Novaes RB, Artioli GG, Freire TO, Coelho DF, Scagliusi FB, Rogeri PS, Roschel H, Ugrinowitsch C, Lancha AH Jr: Effects of creatine supplementation on glucose tolerance and insulin sensitivity in sedentary healthy males undergoing aerobic training. Amino Acids 2008;34:245–250.

14 Curis E, Nicolis I, Moinard C, Osowska S, Zerrouk N, Benazeth S, Cynober L: Almost all about citrulline in mammals. Amino Acids 2005;29:177–205.

15 Kamoun P, Rabier D, Bardet J, Parvy P: Citrulline concentrations in human plasma after arginine load. Clin Chem 1991;37:1287.

16 Mehta S, Stewart DJ, Levy RD: The hypotensive effect of L-arginine is associated with increased expired nitric oxide in humans. Chest 1996;109: 1550–1555.

17 Sureda A, Cordova A, Ferrer MD, Perez G, Tur JA, Pons A: L-Citrulline malate influence over branched chain amino acid utilization during exercise. Eur J Appl Physiol 2010;110:341–351.

18 Romero MJ, Platt DH, Caldwell RB, Caldwell RW: Therapeutic use of citrulline in cardiovascular disease. Cardiovasc Drug Rev 2006;24:275–290.

19 Rouge C, Des Robert C, Robins A, Le Bacquer O, Volteau C, De La Cochetiere MF, Darmaun D: Manipulation of citrulline availability in humans. Am J Physiol Gastrointest Liver Physiol 2007;293: G1061–G1067.

20 Olek RA, Ziemann E, Grzywacz T, Kujach S, Luszczyk M, Antosiewicz J, Laskowski R: A single oral intake of arginine does not affect performance during repeated Wingate anaerobic test. J Sports Med Phys Fitness 2010;50:52–56.

21 Vanhatalo A, Bailey SJ, Blackwell JR, DiMenna FJ, Pavey TG, Wilkerson DP, Benjamin N, Winyard PG, Jones AM: Acute and chronic effects of dietary nitrate supplementation on blood pressure and the physiological responses to moderate-intensity and incremental exercise. Am J Physiol Regul Integr Comp Physiol 2010;299:R1121–R1131.

22 Hartman WJ, Torre PM, Prior RL: Dietary citrulline but not ornithine counteracts dietary arginine deficiency in rats by increasing splanchnic release of citrulline. J Nutr 1994;124:1950–1960.

23 Van de Poll MC, Soeters PB, Deutz NE, Fearon KC, Dejong CH: Renal metabolism of amino acids: its role in interorgan amino acid exchange. Am J Clin Nutr 2004;79:185–197.

24 Curis E, Nicolis I, Moinard C, Osowska S, Zerrouk N, Bénazeth S, Cynober LA: Almost all about citrulline in mammals. Amino Acids 2005;29:177–205.

25 Hickner RC, Tanner CJ, Evans CA, Clark PD, Haddock A, Fortune C, Geddis H, Waugh W, McCammon M: L-Citrulline reduces time to exhaustion and insulin response to a graded exercise test. Med Sci Sports Exerc 2006;38:660–666.

26 Vanuxem D, Duflot JC, Prevot H, Bernasconi P, Blehaut H, Fornaris E, Vanuxem P: Influence of an anti-asthenia agent, citrulline malate, on serum lactate and ammonia kinetics during a maximum exercise test in sedentary subjects. Sémin Hôp Paris 1990;66:477–481.

27 Sureda A, Cordova A, Ferrer MD, Tauler P, Perez G, Tur JA, Pons A: Effects of L-citrulline oral supplementation on polymorphonuclear neutrophils oxidative burst and nitric oxide production after exercise. Free Radic Res 2009;43:828–835.

28 Aguilo A, Tauler P, Sureda A, Cases N, Tur J, Pons A: Antioxidant diet supplementation enhances aerobic performance in amateur sportsmen. J Sports Sci 2007;25:1203–1210.

29 Tidball JG: Inflammatory processes in muscle injury and repair. Am J Physiol Regul Integr Comp Physiol 2005;288:R345–R353.

30 Pérez-Guisado J, Jakeman PM: Citrulline malate enhances athletic anaerobic performance and relieves muscle soreness. J Strength Cond Res 2010;24:1215–1222.

31 Bendahan D, Mattei JP, Ghattas B, Confort-Gouny S, Le Guern ME, Cozzone PJ: Citrulline/malate promotes aerobic energy production in human exercising muscle. Br J Sports Med 2002;36:282–289.

32 Eto K, Suga S, Wakui M, Tsubamoto Y, Terauchi Y, Taka J, Aizawa S, Noda M, Kimura S, Kasai H, Kadowaki T: NADH shuttle system regulates K_{ATP} channel-dependent pathway and steps distal to cytosolic Ca^{2+} concentration elevation in glucose-induced insulin secretion. J Biol Chem 1999;274:25386–25392.

33 Wu JL, Wu QP, Huang JM, Chen R, Cai M, Tan JB: Effects of L-malate on physical stamina and activities of enzymes related to the malate-aspartate shuttle in liver of mice. Physiol Res 2007;56:213–220.

34 Bowtell JL, Bruce M: Glutamine: an anaplerotic precursor. Nutrition 2002;18:222–224.

35 Minuskin ML, Lavine ME, Ulman EA, Fisher H: Nitrogen retention, muscle creatine and orotic acid excretion in traumatized rats fed arginine and glycine-enriched diets. J Nutr 1981;111:1265–1274.

36 Wagenmakers AJ: Muscle amino acid metabolism at rest and during exercise: role in human physiology and metabolism. Exerc Sport Sci Rev 1998;26:287–314.

37 Daussin FN, Zoll J, Ponsot E, Dufour SP, Doutreleau S, Lonsdorfer E, Ventura-Clapier R, Mettauer B, Piquard F, Geny B, Richard R: Training at high exercise intensity promotes qualitative adaptations of mitochondrial function in human skeletal muscle. J Appl Physiol 2008;104:1436–1441.

38 King DE, Mainous AG 3rd, Geesey ME: Variation in L-arginine intake follow demographics and lifestyle factors that may impact cardiovascular disease risk. Nutr Res 2008;28:21–24.

39 Castillo L, Chapman TE, Sanchez M, Yu YM, Burke JF, Ajami AM, Vogt J, Young VR: Plasma arginine and citrulline kinetics in adults given adequate and arginine-free diets. Proc Natl Acad Sci USA 1993;90:7749–7753.

40 Boger RH, Bode-Boger SM: The clinical pharmacology of L-arginine. Annu Rev Pharmacol Toxicol 2001;41:79–99.

41 Sidney M, Morris J: Arginine metabolism: boundaries of our knowledge. 6th Amino Acid Assessment Workshop. J Nutr 2007;137(suppl 2):1602S–1609S.

42 Cynober L: Pharmacokinetics of arginine and related amino acids. J Nutr 2007;137(suppl 2):1646S–1649S.

43 McConell GK: Effects of L-arginine supplementation on exercise metabolism. Curr Opin Clin Nutr Metab Care 2007;10:46–51.

44 Fricke O, Baecker N, Heer M, Tutlewski B, Schoenau E: The effect of L-arginine administration on muscle force and power in postmenopausal women. Clin Physiol Funct Imaging 2008;28:307–311.

Arginine and Citrulline Supplementation in Sports and Exercise: Ergogenic Nutrients?

27

45 Koppo K, Taes YE, Pottier A, Boone J, Bouckaert J, Derave W: Dietary arginine supplementation speeds pulmonary V_{O_2} kinetics during cycle exercise. Med Sci Sports Exerc 2009;41:1626–1632.

46 Liu TH, Wu CL, Chiang CW, Lo YW, Tseng HF, Chang CK: No effect of short-term arginine supplementation on nitric oxide production, metabolism and performance in intermittent exercise in athletes. J Nutr Biochem 2009;20:462–468.

47 Sunderland KL, Greer F, Morales J: V_{O_2} max and ventilatory threshold of trained cyclists are not affected by 28-day L-arginine supplementation. J Strength Cond Res 2011;25:833–837.

48 Tsai PH, Tang TK, Juang CL, Chen KW, Chi CA, Hsu MC: Effects of arginine supplementation on post-exercise metabolic responses. Chin J Physiol 2009;52:136–142.

49 McConell GK, Huynh NN, Lee-Young RS, Canny BJ, Wadley GD: L-Arginine infusion increases glucose clearance during prolonged exercise in humans. Am J Physiol Endocrinol Metab 2006;290:E60–E66.

50 Schaefer A, Piquard F, Geny B, Doutreleau S, Lampert E, Mettauer B, Lonsdorfer J: L-Arginine reduces exercise-induced increase in plasma lactate and ammonia. Int J Sports Med 2002;23:403–407.

51 Wu G, Morris SM Jr: Arginine metabolism: nitric oxide and beyond. Biochem J 1998;336:1–17.

52 Fang YZ, Yang S, Wu G: Free radicals, antioxidants, nutrition. Nutrition 2002;18:872–879.

53 Bassit RA, Curi R, Costa Rosa LF: Creatine supplementation reduces plasma levels of pro-inflammatory cytokines and PGE_2 after a half-ironman competition. Amino Acids 2008;35:425–431.

54 Little JP, Forbes SC, Candow DG, Cornish SM, Chilibeck PD: Creatine, arginine α-ketoglutarate, amino acids, and medium-chain triglycerides and endurance and performance. Int J Sport Nutr Exerc Metab 2008;18:493–508.

55 Paddon-Jones D, Borsheim E, Wolfe RR: Potential ergogenic effects of arginine and creatine supplementation. J Nutr 2004;134(suppl):2888S–2895S.

56 Volek JS, Mazzetti SA, Farquhar WB, Barnes BR, Gomez AL, Kraemer WJ: Physiological responses to short-term exercise in the heat after creatine loading. Med Sci Sports Exerc 2001;33:1101–1108.

57 Camic CL, Housh TJ, Mielke M, Zuniga JM, Hendrix CR, Johnson GO, Schmidt RJ, Housh DJ: The effects of 4 weeks of an arginine-based supplement on the gas exchange threshold and peak oxygen uptake. Appl Physiol Nutr Metab 2010;35:286–293.

58 Chen S, Kim W, Henning SM, Carpenter CL, Li Z: Arginine and antioxidant supplement on performance in elderly male cyclists: a randomized controlled trial. J Int Soc Sports Nutr 2010;7:13.

59 Tang JE, Lysecki PJ, Manolakos JJ, MacDonald MJ, Tarnopolsky MA, Phillips SM: Bolus arginine supplementation affects neither muscle blood flow nor muscle protein synthesis in young men at rest or after resistance exercise. J Nutr 2011;141:195–200.

60 Campbell B, Roberts M, Kerksick C, Wilborn C, Marcello B, Taylor L, Nassar E, Leutholtz B, Bowden R, Rasmussen C, Greenwood M, Kreider R: Pharmacokinetics, safety, effects on exercise performance of L-arginine α-ketoglutarate in trained adult men. Nutrition 2006;22:872–881.

61 Buford BN, Koch AJ: Glycine-arginine-α-ketoisocaproic acid improves performance of repeated cycling sprints. Med Sci Sports Exerc 2004;36:583–587.

62 Stevens BR, Godfrey MD, Kaminski TW, Braith RW: High-intensity dynamic human muscle performance enhanced by a metabolic intervention. Med Sci Sports Exerc 2000;32:2102–2108.

63 Colombani PC, Bitzi R, Frey-Rindova P, Frey W, Arnold M, Langhans W, Wenk C: Chronic arginine aspartate supplementation in runners reduces total plasma amino acid level at rest and during a marathon run. Eur J Nutr 1999;38:263–270.

64 Abel T, Knechtle B, Perret C, Eser P, von Arx P, Knecht H: Influence of chronic supplementation of arginine aspartate in endurance athletes on performance and substrate metabolism – a randomized, double-blind, placebo-controlled study. Int J Sports Med 2005;26:344–349.

65 Jones AM, Carter H: The effect of endurance training on parameters of aerobic fitness. Sports Med 2000;29:373–386.

66 Hauk JM, Hosey RG: Nitric oxide therapy: fact or fiction? Curr Sports Med Rep 2006;5:199–202.

67 Denis C, Dormois D, Linossier MT, Eychenne JL, Hauseux P, Lacour JR: Effect of arginine aspartate on the exercise-induced hyperammonemia in humans: a two periods cross-over trial. Arch Int Physiol Biochim Biophys 1991;99:123–127.

68 Eto B, Peres G, Le Moel G: Effects of an ingested glutamate arginine salt on ammonemia during and after long-lasting cycling. Arch Int Physiol Biochim Biophys 1994;102:161–162.

Dr. Antoni Pons
Laboratori de Ciències de l'Activitat Física, Universitat de les Illes Balears
Crtra Valldemossa, Km 7,5
ES–07122 Palma de Mallorca, Illes Balears (Spain)
Tel. +34 971173171, E-Mail antonipons@uib.es

Lamprecht M (ed): Acute Topics in Sport Nutrition.
Med Sport Sci. Basel, Karger, 2013, vol 59, pp 29–35

Dietary Nitrate and O₂ Consumption during Exercise

Andrew M. Jones · Stephen J. Bailey · Anni Vanhatalo

Sport and Health Sciences, College of Life and Environmental Sciences, St. Luke's Campus,
University of Exeter, Exeter, UK

Abstract

Recent studies have investigated the influence of dietary nitrate supplementation on the physiological responses to exercise. Specifically, it has been reported that enhancing nitric oxide (NO) bioavailability through supplementation of the diet with nitrate salts or nitrate-rich beetroot juice reduces the O_2 cost of exercise and improves exercise performance. The lower O_2 cost for a given sub-maximal work rate following nitrate ingestion indicates that muscle efficiency is enhanced either as a consequence of a reduced energy cost of contraction or enhanced mitochondrial efficiency. The positive effects of nitrate supplementation on the O_2 cost of sub-maximal exercise can be manifested acutely (i.e. 2.5 h following ingestion) and maintained for at least 15 days if supplementation is continued. Most recently, the influence of dietary nitrate supplementation on time trial performance in competitive cyclists has been investigated. Studies have shown a 1–2% reduction in the time to complete time trial distances between 4 and 16 km. The dose of nitrate that has been shown to improve exercise efficiency can readily be achieved through the consumption of 0.5 litre of beetroot juice or an equivalent high-nitrate foodstuff. Following a 5- to 6-mmol bolus of nitrate, plasma [nitrite] typically peaks within 2–3 h and remains elevated for a further 6–9 h before declining towards baseline. Therefore, consuming nitrate approximately 3 h prior to competition or training is recommended if athletes wish to explore the ergogenic potential of nitrate supplementation.

Nitric oxide (NO) may modulate skeletal muscle function through its role in the regulation of blood flow, muscle contractility, glucose and calcium homeostasis, and mitochondrial respiration and biogenesis. NO was originally believed to be generated solely through the oxidation of L-arginine in a reaction catalysed by nitric oxide synthase (NOS), with nitrite (NO_2^-) and nitrate (NO_3^-) considered to be inert by-products of this process. However, it is now clear that NO_3^- and NO_2^- can be recycled back into NO under certain physiological conditions. This reduction of NO_3^- to NO_2^- and subsequently of NO_2^- to NO may be important as a means to increase NO production when NO synthesis by the NOS enzymes is impaired and in conditions of low O_2 availability, such as may occur in skeletal muscle during exercise.

In addition to being produced endogenously through the arginine-NOS pathway, plasma concentrations of NO_3^- and NO_2^- can be increased by dietary means. Green leafy vegetables such as lettuce, spinach, rocket, celery and beetroot are particularly rich in nitrate. Therefore, dietary nitrate supplementation represents a practical method to increase plasma $[NO_2^-]$ with this serving as a circulating 'reservoir' for NO production. Nitrate can be ingested either as a salt such as potassium or sodium nitrate [1, 2], or through the consumption of nitrate-rich whole vegetables or vegetable products such as beetroot juice [3–6]. Given the importance of NO in vascular and metabolic control, there are reasons to believe that augmenting NO bioavailability might be important in optimising skeletal muscle function during exercise. Indeed, recent evidence indicates that elevating plasma $[NO_2^-]$ through dietary nitrate supplementation is associated with enhanced muscle efficiency, fatigue resistance and performance.

Nitrate and O_2 Uptake

The first study to suggest a relationship between nitrate ingestion and changes in exercise metabolism was published in 2007 by Larsen et al. [1]. These authors reported that 3 days of sodium nitrate supplementation increased plasma $[NO_2^-]$ and reduced the O_2 cost of sub-maximal cycle exercise by 3–4%. These findings were surprising because it is well established that the O_2 cost of exercising at a given sub-maximal power output is highly predictable. During cycle ergometry, for example, it is expected that pulmonary O_2 uptake ($\dot{V}O_2$) will increase by approximately 10 ml/min for every additional watt of external power output, i.e. the functional 'gain' is ~10 $ml \cdot min^{-1} \cdot W^{-1}$. The findings of Larsen et al. [1] were corroborated in the study of Bailey et al. [3] in which nitrate was administered in the form of beetroot juice. Following 3–6 days of beetroot juice supplementation (0.5 l/day), the plasma $[NO_2^-]$ was doubled, the steady-state $\dot{V}O_2$ during moderate-intensity exercise was reduced and the $\dot{V}O_2$ 'slow component' during severe-intensity exercise was attenuated (fig. 1). Collectively, these results suggested that a short-term, natural dietary intervention improved the efficiency of muscular work.

In the studies of Larsen et al. [1] and Bailey et al. [3], nitrate supplementation was continued for 3–6 days. However, a similar reduction in steady-state $\dot{V}O_2$ during moderate-intensity cycle ergometry has been reported following acute nitrate supplementation by Vanhatalo et al. [5]. These authors reported a significant reduction in steady-state $\dot{V}O_2$ just 2.5 h following beetroot juice ingestion, an effect that was maintained when supplementation was continued for up to 15 days. Subsequent studies showed that a reduction in steady-state $\dot{V}O_2$ following beetroot juice consumption was also evident during two-legged knee-extensor exercise [4] and treadmill running [7].

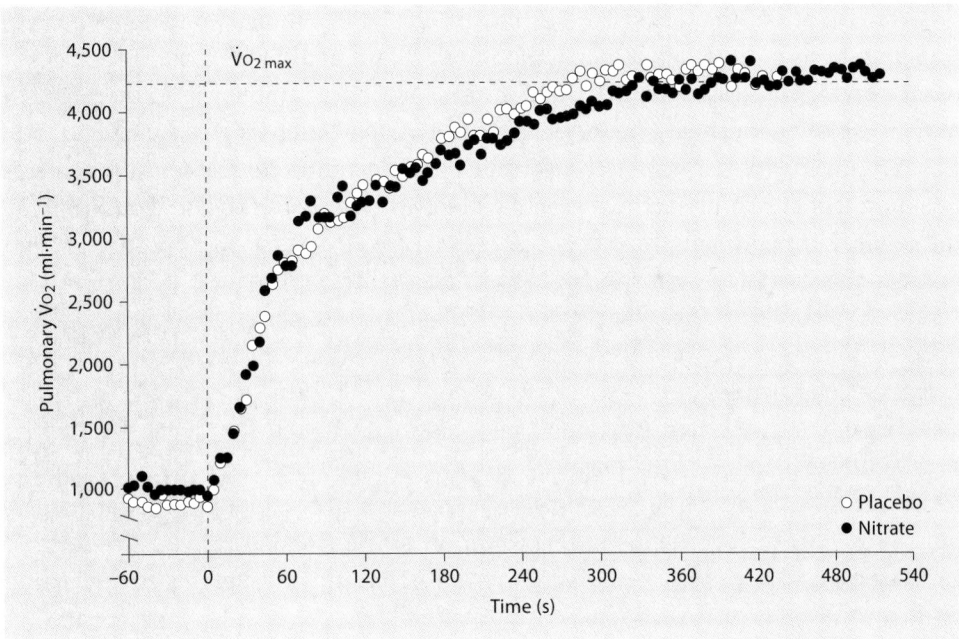

Fig. 1. Influence of dietary nitrate supplementation with beetroot juice on O_2 uptake during severe-intensity exercise in a representative subject. The $\dot{V}O_2$ slow component is attenuated following nitrate supplementation, delaying the attainment of $\dot{V}O_{2\,max}$ and increasing the time-to-exhaustion.

Mechanistic Bases for Physiological Effects of Nitrate

Theoretically, a lower O_2 cost during exercise at the same power output could result from: (1) a lower ATP cost of muscle contraction for the same force production (i.e. improved muscle contractile efficiency), and/or (2) a lower O_2 consumption for the same rate of oxidative ATP resynthesis (i.e. improved mitochondrial efficiency).

Bailey et al. [4] used ^{31}P-magnetic resonance spectroscopy to assess absolute muscle concentration changes in phosphocreatine ([PCr]), inorganic phosphate ([P_i]), and adenosine diphosphate ([ADP]), as well as pH, following nitrate supplementation. The estimated ATP turnover rates from PCr hydrolysis and oxidative phosphorylation were lower following 6 days of beetroot juice supplementation, such that there was a significant reduction in the estimated total ATP turnover rate during both low- and high-intensity exercise [4]. The authors suggested that elevated NO production following nitrate supplementation may have reduced skeletal muscle ATP turnover by reducing the activity of actomyosin ATPase and/or Ca^{2+}-ATPase. The smaller changes in [ADP], [P_i] and [PCr] following NO_3^- supplementation would be predicted to reduce the stimuli for increasing oxidative phosphorylation [8].

The second possibility, that nitrate supplementation enhances mitochondrial efficiency, was recently investigated by Larsen et al. [9]. These authors isolated mitochondria from the vastus lateralis muscle of healthy humans supplemented with sodium nitrate. Nitrate supplementation reduced proton leakage and uncoupled respiration, which increased the mitochondrial P/O ratio (the amount of ATP produced/ oxygen used). Interestingly, the increased P/O ratio following nitrate supplementation was correlated with the reduction in whole body $\dot{V}o_2$ during exercise [9]. It appears therefore that nitrate supplementation may improve exercise efficiency by improving the efficiency of both muscle contraction (reduced ATP cost of force production) and mitochondrial oxidative phosphorylation (increased P/O ratio).

Exercise Performance

In the study of Bailey et al. [3], plasma $[NO_2^-]$ was doubled and high-intensity exercise tolerance was enhanced by 16% following several days of beetroot juice supplementation. Subsequent experiments have reported improvements in exercise tolerance of 25% during two-legged knee-extensor exercise [4] and 15% during treadmill running [7] following 6 days of beetroot juice supplementation.

It is well documented that exercise performance is compromised in a hypoxic environment relative to normoxia (21% O_2). It is noteworthy, therefore, that Vanhatalo et al. [10] reported that nitrate supplementation with beetroot juice restored muscle performance in hypoxia (14% inspired O_2; equivalent to 4,000 m altitude) to that observed in normoxia. Specifically, in hypoxia, nitrate supplementation resulted in a 20% extension of the time-to-exhaustion during high-intensity knee-extensor exercise. Vanhatalo et al. [10] also reported that nitrate supplementation improved muscle oxidative function in hypoxia, suggesting that muscle oxygenation may have been enhanced. Consistent with this interpretation, Kenjale et al. [11] reported that beetroot juice supplementation resulted in a 17–18% longer time to claudication pain and peak walking time during incremental exercise in patients with peripheral arterial disease. The authors attributed these effects to NO_2^--related improvement in peripheral tissue oxygenation. Collectively, these results have potential performance implications for athletes competing at altitude and for improving functional capacity in clinical conditions where tissue O_2 supply may be compromised.

The improvement in exercise *tolerance* at a given power output following nitrate supplementation has been reported to be ~16–25% [3, 4, 7]. However, the magnitude of improvement in exercise *performance* would be expected to be smaller. For example, a ~20% improvement in time-to-exhaustion would be expected to correspond to an improvement in exercise performance (time taken to cover a set distance) of 1–2% [12]. To address the potential 'real-world' benefit of nitrate supplementation, Lansley et al. [13] asked competitive but sub-élite cyclists to complete, on separate days, 4.0- and 16.1-km time trials, following acute beetroot juice or placebo ingestion. Nitrate improved 4.0- and 16.1-km time trial performance by ~2.7% compared to the placebo conditions (fig. 2). These improvements in exercise performance were consequent to

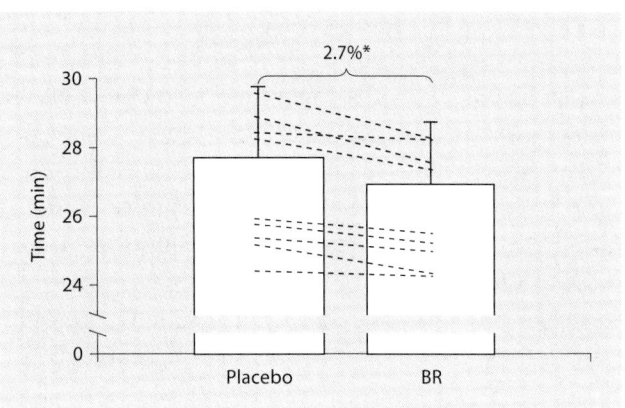

Fig. 2. The time to complete a 16.1-km cycle time trial was reduced by approximately 2.7% following nitrate supplementation in trained cyclists. The bars represent the group mean ± SEM whereas the dashed lines represent the individual responses (n = 9). BR = Beetroot juice. *Statistically significant (p < 0.05).

the maintenance of a higher mean power output and an increase in the power output/$\dot{V}O_2$ ratio. The results showed that trained subjects were able to produce a higher power output for the same oxidative energy turnover, resulting in an improved exercise performance following nitrate supplementation. Improved cycle time trial performance following nitrate supplementation has also been reported by Cermak et al. [14]. These authors reported that 6 days of beetroot juice supplementation (8 mmol/day) significantly reduced $\dot{V}O_2$ at two sub-maximal work rates and improved mean power output and 10-km time trial performance (by 1.2%) in trained cyclists.

Despite these positive results with 'sub-élite' athletes, it remains unclear whether nitrate supplementation might enhance performance in athletes of the highest calibre. Wilkerson et al. [15] reported that acute nitrate supplementation did not enhance 50-mile time trial performance in a group of well-trained cyclists, but also found a significant correlation (r = –0.83) between the increase in plasma [NO_2^-] and the improvement in time trial performance. These results suggest that the nitrate dosing regimen (i.e. amount and timing of ingestion) may be critical. It should also be considered that highly-trained subjects are likely to have: (1) higher NOS activity such that the nitrate-nitrite-NO pathway may be relatively less important for the generation of NO, and (2) greater mitochondrial and capillary density which might limit the development of hypoxia and acidosis in skeletal muscle during exercise, preserving NOS function and reducing the requirement for nitrite reduction to NO. Further research is needed to elucidate the influence of NO_3^- supplementation on exercise efficiency in elite athletes.

Conclusion

Dietary nitrate appears to hold promise as a natural means to enhance NO bioavailability. Dietary nitrate supplementation reduces resting blood pressure and may therefore be important in maintaining and promoting cardiovascular health. It is now well established that acute and chronic nitrate supplementation can reduce the O_2 cost of sub-maximal exercise. This improvement in muscular efficiency may be linked to a reduced energy cost of muscle contraction and/or to enhanced efficiency of mitochondrial ATP production. Several studies indicate that, at least in recreational or moderately-trained subjects, nitrate supplementation can extend exercise tolerance and improve time trial performance. Additional work is required before the effectiveness of nitrate supplementation on performance in different types of physical activity and in different human populations is fully understood.

Disclosure Statement

The authors have no conflicts of interest to disclose.

References

1 Larsen FJ, Weitzberg E, Lundberg JO, Ekblom B: Effects of dietary nitrate on oxygen cost during exercise. Acta Physiol 2007;191:59–66.
2 Larsen FJ, Weitzberg E, Lundberg JO, Ekblom B: Dietary nitrate reduces maximal oxygen consumption while maintaining work performance in maximal exercise. Free Radic Biol Med 2010;48: 342–347.
3 Bailey SJ, Winyard P, Vanhatalo A, Blackwell JR, DiMenna FJ, Wilkerson DP, Tarr J, Benjamin N, Jones AM: Dietary nitrate supplementation reduces the O_2 cost of low-intensity exercise and enhances tolerance to high-intensity exercise in humans. J Appl Physiol 2009;107:1144–1155.
4 Bailey SJ, Fulford J, Vanhatalo A, Winyard P, Blackwell JR, DiMenna FJ, Wilkerson DP, Benjamin N, Jones AM: Dietary nitrate supplementation enhances muscle contractile efficiency during knee-extensor exercise in humans. J Appl Physiol 2010; 109:135–148.
5 Vanhatalo A, Bailey SJ, Blackwell JR, DiMenna FJ, Pavey TG, Wilkerson DP, Benjamin N, Winyard PG, Jones AM: Acute and chronic effects of dietary nitrate supplementation on blood pressure and the physiological responses to moderate-intensity and incremental exercise. Am J Physiol Regul Integr Comp Physiol 2010;299:R1121–1131.
6 Webb AJ, Patel N, Loukogeorgakis S, Okorie M, Aboud Z, Misra S, Rashid R, Miall P, Deanfield J, Benjamin N, MacAllister R, Hobbs AJ, Ahluwalia A: Acute blood pressure lowering, vasoprotective, and antiplatelet properties of dietary nitrate via bioconversion to nitrite. Hypertension 2008;51:784–790.
7 Lansley KE, Winyard PG, Fulford J, Vanhatalo A, Bailey SJ, Blackwell JR, DiMenna FJ, Gilchrist M, Benjamin N, Jones AM: Dietary nitrate supplementation reduces the O_2 cost of walking and running: a placebo-controlled study. J Appl Physiol 2011;110: 591–600.
8 Mahler M: First-order kinetics of muscle oxygen consumption, and equivalent proportionality between Qo_2 and phosphorylcreatine level. Implications for the control of respiration. J Gen Physiol 1985;86:135–165.
9 Larsen FJ, Schiffer TA, Borniquel S, Sahlin K, Ekblom B, Lundberg JO, Weitzberg E: Dietary inorganic nitrate improves mitochondrial efficiency in humans. Cell Metab 2011;13:149–159.
10 Vanhatalo A, Fulford J, Bailey SJ, Blackwell JR, Winyard PG, Jones AM: Dietary nitrate reduces muscle metabolic perturbation and improves exercise tolerance in hypoxia. J Physiol 2011;589:5517–5528.
11 Kenjale AA, Ham KL, Stabler T, Robbins JL, Johnson JL, Vanbruggen M, Privette G, Yim E, Kraus WE, Allen JD: Dietary nitrate supplementation enhances exercise performance in peripheral arterial disease. J Appl Physiol 2011;110:1582–1591.

12 Hopkins WG, Hawley JA, Burke LM: Design and analysis of research on sport performance enhancement. Med Sci Sports Exerc 1999;31:472–485.

13 Lansley KE, Winyard PG, Bailey SJ, Vanhatalo A, Wilkerson DP, Blackwell JR, Gilchrist M, Benjamin N, Jones AM: Acute dietary nitrate supplementation improves cycling time trial performance. Med Sci Sports Exerc 2011;43:1125–1131.

14 Cermak NM, Gibala MJ, van Loon LJ: Nitrate supplementation's improvement of 10-km time-trial performance in trained cyclists. Int J Sport Nutr Exerc Metab 2012;22:64–71.

15 Wilkerson DP, Hayward GM, Bailey SJ, Vanhatalo A, Blackwell JR, Jones AM: Influence of acute dietary nitrate supplementation on 50 mile time trial performance in well-trained cyclists. Eur J Appl Physiol 2012;Apr 20, in press.

Andrew M. Jones, PhD, Prof.
Sport and Health Sciences, University of Exeter
Heavitree Road, Exeter EX1 2LU (UK)
E-Mail a.m.jones@exeter.ac.uk

Lamprecht M (ed): Acute Topics in Sport Nutrition.
Med Sport Sci. Basel, Karger, 2013, vol 59, pp 36–46

GABA Supplementation and Growth Hormone Response

Michael Powers

Marist College, Poughkeepsie, N.Y., USA

Abstract

The secretion of growth hormone (GH) is regulated through a complex neuroendocrine control system, especially by the functional interplay of two hypothalamic hormones, GH-releasing hormone and somatostatin. These hormones are subject to modulation by a host of neurotransmitters and are the final mediators of endocrine and neural influences for GH secretion. Interest in the possible role of γ-aminobutyric acid (GABA) in the control of GH secretion began decades ago. However, interest in its role as an ergogenic aid is only recent. It is well accepted that GABAergic neurons are found in the hypothalamus and recent evidence suggests its secretion within the pituitary itself. Inhibition of GABA degradation and blockade of GABA transmission as well as administration of GABA and GABA mimetic drugs have all been shown to affect GH secretion. However, there are many controversial findings. The effects may depend on the site of action within the hypothalamic-pituitary unit and the hormonal milieu. Experimental and clinical evidence support the presence of a dual action of GABA – one mediated centrally, the other exerted directly at the pituitary level. The two sites of action may be responsible for excitatory and inhibitory effects of GABA on GH secretion. This chapter will outline the anatomical basis for possible influences of GABA on GH secretion and present evidence for a role of GABA in the control of GH release by actions at either hypothalamic or pituitary sites. The potential ergogenic benefits of oral GABA supplementation will also be discussed.

Growth Hormone

Growth hormone (GH) is a peptide hormone that plays an important role in the growth and maintenance of skeletal muscle, stimulating increases in muscle and cartilage protein synthesis, fatty acid use, and cellular amino acid uptake [1]. Secreted by somatotropic cells of the anterior pituitary, it exhibits a great deal of molecular heterogeneity and circulates in multiple forms, only some of which are biologically active [2]. This has physiological significance, as the different forms have been shown to have different effects on lipid, carbohydrate and protein metabolism. In the adult human, approximately five pulses of GH are secreted during a 24-hour

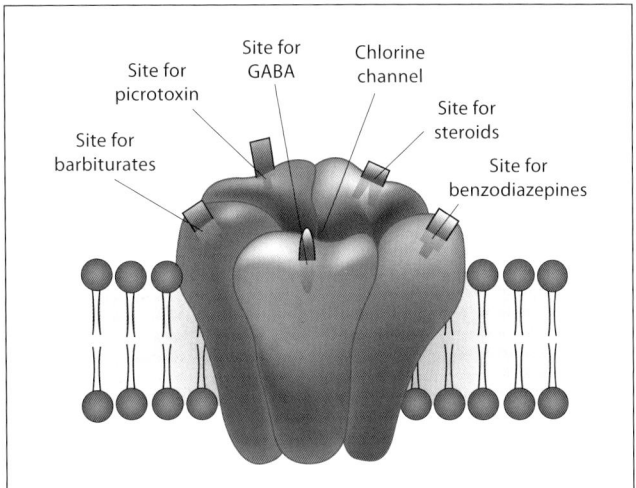

Fig. 1. The GABA$_A$ ligand-gated chloride ion channel receptor complex [from http://thebrain.mcgill.ca/flash/i/i_04/i_04_m/i_04_m_peu/i_04_m_peu.html].

period with a larger peak occurring at the onset of sleep at night [3]. However, several stimuli, including exercise [4] and amino acid administration [5] purportedly alter this pattern. GH secretion is regulated through a complex neuroendocrine control system that includes two main hypothalamic regulators, GH-releasing hormone (GHRH) and somatostatin (SS), exerting stimulatory and inhibitory influences, respectively, on somatotropic cells. Until recently, GH secretion was thought to be determined simply by the balance between these two hypothalamic peptides. It is now accepted that GH secretion is also influenced by other peptides produced in the periphery such as ghrelin [6]. Regardless of the stimulus, GHRH and SS are considered the final mediators of pituitary GH secretion and are subject to modulation by other hypothalamic peptides and by complex networks of neurons and neurotransmitters.

γ-Aminobutyric Acid

γ-Aminobutyric acid (GABA) is the chief inhibitory neurotransmitter in the central nervous system. Endogenously, it is synthesized from the decarboxylation of glutamate by the enzyme glutamic acid decarboxylase (GAD) and is metabolized to succinate by the sequential actions of GABA transaminase and succinic semialdehyde dehydrogenase. Neurons that secrete GABA are referred to as GABAergic neurons and have chiefly inhibitory actions, as receptor binding at pre- and postsynaptic membranes results in the opening or closing of ion channels. Two general classes of GABA receptors have been identified – GABA$_A$ (fig. 1) in which the receptor is part of a ligand-gated chloride ion channel complex [7], and GABA$_B$ metabotropic receptors [8]. The GABA$_B$ receptors are guanine-nucleotide-binding proteins coupled to second messenger generating systems that open potassium channels. Thus, receptor binding generally results in the influx of negatively charged chloride ions or the efflux of positively charged potassium. GABA$_B$ receptor binding can also decrease the cell's conductance to calcium. More recently, a third GABA receptor has been suggested and labeled as GABA$_C$ [9]. Like GABA$_A$, it is part of a ligand-gated chloride channel complex. However, some simply consider it a subtype of the GABA$_A$ receptor complex, as a number of GABA$_A$ and GABA$_B$ subtypes have been identified [7, 8]. Once secreted, the synaptic actions of GABA are terminated by degrading enzymes and the high-affinity uptake systems (GABA transporters) of glial cells and neurons that can repackage it into vesicles [10].

Multiple feedback mechanisms control GABA concentration [11]. It is most highly concentrated in the substantia nigra and globus pallidus of the basal ganglia, followed by the hypothalamus, the periaqueductal grey matter and the hippocampus [12]. There are several ways of increasing GABAergic activity in the human brain and a number of GABA agonists have been identified, such as muscimol, benzodiazepines (e.g. diazepam), barbiturates (e.g. phenobarbital), propofol and progabid. These directly increase inhibitory chloride conductance or upregulate the effect of synaptic released GABA on the GABA$_A$ receptor. Muscimol binds to the same site on the GABA$_A$ receptor complex as GABA itself, while benzodiazepines and barbiturates bind to separate regulatory sites on the receptor complex. Thus, muscimol can enhance chloride conductance independent of GABA, while the other drugs only affect the efficacy and potency of GABA once it binds to the receptor. Progabid is also considered a GABA$_B$ agonist, while the drug baclofen is a selective GABA$_B$ agonist. Other drug types include GABA transporter blockers, which prolong the action of GABA in the synaptic cleft by inhibiting its uptake. Vigabatrin and sodium valproate are drugs that inhibit GABA transaminase and slow the degradation of GABA. Valproate is also thought to stimulate GABA synthesis. Both mechanisms, increasing synthesis or decreasing degradation, would increase intra- and extracellular GABA concentrations.

GABA and Pituitary Function

There is more than adequate anatomical evidence that GABA is involved in the regulation of many pituitary hormones both centrally and at the level of the gland itself. Inhibition of GABA degradation and blockade of GABA transmission as well as administration of GABA and GABA mimetic drugs have all been shown to affect pituitary hormone secretion [13–31]. While not completely understood, GABA may act at different sites within the hypothalamic-pituitary unit and these actions may also depend on the hormonal milieu. There is a remarkable density of GABAergic endings in the arcuate and periventricular nuclei and these endings are in synaptic contact with other hypothalamic cells, including those that produce dopamine and releasing and inhibiting regulatory hormones [32, 33]. A vast majority of neurons from these nuclei terminate in the median eminence where they release regulatory hormones into the hypophyseal portal blood. They express multiple receptors for different neurotransmitters and their location in the median eminence brings them in contact with a host of other neurotransmitters and nerve endings [34–36]. Collectively they are referred to as the tuberoinfundibular system, which provides a major pathway for the release of GHRH, SS and other compounds and their influence on pituitary function. There are GABAergic fibers that also project from the arcuate and periventricular nuclei and terminate in the median eminence. Thus, the appropriate secretion of releasing and inhibitory compounds, such as GHRH and SS, can be regulated by different neurotransmitters within the nuclei and within the median eminence, including GABA [34–36]. This supports the central role of GABA and its influence on hypothalamic-pituitary function.

In addition to GABA's central effect, there is sufficient evidence for its direct influence on pituitary cells, as GABA receptors have been identified on gonadotropes, corticotropes and somatotropes and fairly high concentrations of GABA have been measured within the gland itself [16, 35, 37]. The physiological sources of pituitary GABA are not completely established, but it is generally accepted that it is synthesized by hypothalamic neurons and reaches the pituitary by two possible routes. The first involves secretion of GABA into the hypophyseal portal system and the second involves direct innervation of endocrine cells within the intermediate lobe. As mentioned above, the tuberoinfundibular pathway contains GABAergic fibers that terminate at the median eminence. Like dopamine, GHRH and SS, GABA secreted from these nerve endings can be released into the portal blood and transported to the anterior pituitary. This theory is supported by elevations in portal blood GABA observed following electrical stimulation of the median eminence [38] and inhibition of GAD [23]. These findings coupled with the presence of relatively high concentrations of GABA observed within the anterior pituitary in the absence of GAD [35], strongly suggest that GABA is indeed secreted into the hypophyseal portal vessels. Additionally, the number of GABA-binding sites within the anterior pituitary has been shown to increase following lesions of the median eminence [39]. This suggests an upregulation of the receptors following the elimination of the GABA which normally reached the receptors via the portal blood. Thus, it is likely that the hypophyseal portal vessels are the primary source of pituitary GABA. However, hypothalamic GABAergic axons can also directly reach and terminate on endocrine cells of the intermediate lobe [40] and, more recently, evidence suggests that GABA can actually be produced and stored within the pituitary gland itself [41]. The endocrine cells were also found to express $GABA_A$ and $GABA_B$ receptor subunits. Taken together, the data strongly imply the existence of novel autocrine and paracrine modes of regulation of pituitary function by GABA (fig. 2).

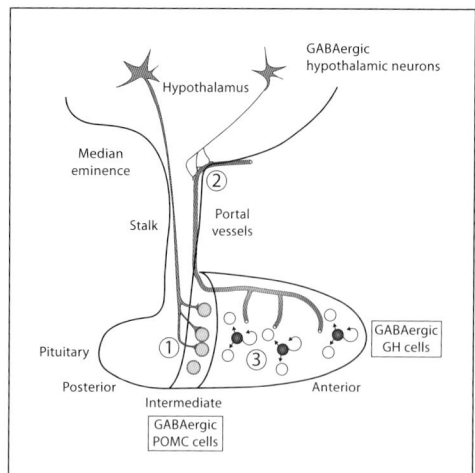

Fig. 2. Neuronal ((1)), neurohemal ((2)), and novel para/autocrine ((3)) control of pituitary function by GABA [from 41].

GABA and Growth Hormone Secretion

The actual role of GABA on GH secretion has been a source of considerable controversy, as numerous mechanisms of action have been proposed and a dual effect appears to exist. As mentioned above, GABA can act centrally by interfering with the activity of other neurotransmitters associated with GH secretion like dopamine [18, 24]. The centrally acting dopamine antagonist pimozide has been shown to blunt the GH response to GABA [18]. This was not observed when the peripheral dopamine antagonist domperidone was administered. However, the GH response to GABA was not completely eliminated by pimozide, suggesting that other mechanisms participate in the neuroendocrine effect of GABA. Early studies utilizing an intraventricular method of GABA administration indicate that it exerts a dose-related stimulatory effect on GH secretion [29, 30]. The effects were blocked by the GABA$_A$ antagonist bicuculline [30]. In a similar study, Willoughby et al. [31] observed an immediate increase in GH secretion when muscimol was injected directly into the periventricular area. Likewise, Spencer et al. [27] observed a rapid increase in plasma GH following intraventricular administration of GABA (10 mg). However, administration of 100 mg was inhibitory and decreased GH secretion. A plausible explanation for these observations would be the inhibition of SS release by GABA [31, 33, 42]. Not all studies are in agreement however [43].

In addition to central action, there is evidence that GABA can act directly on pituitary somatotropes [14, 16]. For example, stimulation of GH secretion has been observed in vitro following infusion of GABA directly into the pituitary [16]. The effect was purely stimulatory and transient, peaking at approximately 4 min and lasting approximately 20 min. Increases were also observed following muscimol administration. The response to muscimol was reduced when bicuculline was administered, while baclofen had no effect on GH secretion. Additionally, the addition of benzodiazepine and secobarbital (barbiturate), which are known to potentiate the GABAergic response, enhanced the GH response to muscimol. Using a similar technique, Acs et al. [14] observed a dose-dependent increase in GH secretion following GABA infusion. Administration of a $GABA_A$ channel blocker diminished the GH response to GABA by 60%. A desensitization of the receptors was also suggested as a gradual decrease in GH secretion was observed following prolonged stimulation with GABA. Similar results have been reported elsewhere [15].

Inhibitory actions by GABA on GH have also been described and appear to occur centrally [20]. Diazepam has been shown to inhibit GH secretion by inhibition of dopaminergic transmission [24]. Likewise, intravenous administration of muscimol has been shown to inhibit secretory peaks of plasma GH [22]. In that study, GH inhibition was also observed when brain GABA levels were increased by injecting γ-acetylenic-GABA. Conversely, an intravenous injection of bicuculline triggered an early rise in plasma GH. It is plausible that GABA inhibits spontaneous GH release by inhibiting the secretion of GHRH [17]. It has been suggested that this dual effect by GABA on GH secretion might be explained by the location of action. Fiók et al. [22] reported poor penetration of GABA into the brain parenchyma following intraventricular injection. They suggested that GABA, when given intraventricularly, suppresses the activity of SSergic neurons in the periventricular region thereby leading to an elevation in GH secretion. On the other hand, it has been suggested that when GABA mimetic drugs are given peripherally, they reach the site of origin of GHRH neurons in the arcuate nuclear region and inhibit the release of GHRH, resulting in a fall in plasma GH.

GABA Supplementation

GABA is commercially available in its synthetic form as a nutritional supplement and is marketed as an anabolic agent via its ability to enhance endogenous GH production. There is evidence for this claim as both subcutaneous [13] and intravenous [27] administration of GABA has been shown to enhance GH secretion. The elevations following subcutaneous injection occurred in a dose-dependent fashion with the peak response occurring 20 min after injection. Administration of GHRH antibody did not interfere with the GH secretion, suggesting that GHRH has only a minor, if any role in GABA-induced GH secretion, at least when administered peripherally. It must be noted that both of these studies used an animal model and a parenteral route of administration. While studies investigating oral administration in humans are limited, the results favor GABA supplementation. Oral administration of sodium valproate [28] and baclofen [25] has been shown to stimulate GH secretion. Likewise, significant elevations of plasma GH have been observed following a single 5 g oral

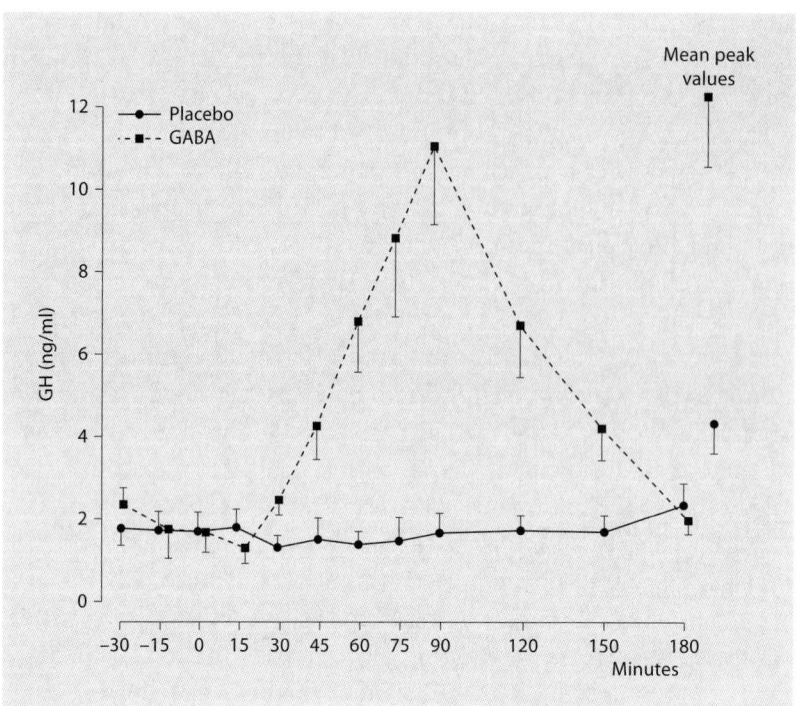

Fig. 3. Resting plasma GH pattern after oral administration of 5 g GABA or placebo [from 19].

dose of GABA [19]. The acute increase was observed in all 19 of the subjects studied (fig. 3). However, administration of 18 g GABA daily for 4 days by 8 additional subjects caused a significant blunting of the overall GH response to insulin hypoglycemia. In a similar study, Cavignini et al. [18] observed a significant elevation in GH following oral administration of 5 g GABA. More recently, 3 g of oral GABA has been shown to increase GH secretion [26]. In that study, an augmentation of the resistance exercise-induced GH response was also observed (fig. 4a, b). However, conflicting results regarding the response during cycling exercise have been reported when valproate is ingested [21, 28]. Steardo et al. [28] observed a significant increase in resting GH concentrations following valproate ingestion, while the GH response to a 20-min bout of cycling was markedly inhibited by it (fig. 5). To the author's knowledge, these are the only studies to investigate the effects of GABA during exercise, with only one investigating true GABA administration [26].

Peripherally administered GABA does not easily cross the blood-brain barrier and the perikarya of hypothalamic secretory neurons are localized in areas protected by this barrier. However, the axon terminals of these neurons are localized in the median eminence, which lacks the blood-brain barrier. It is known that intravenous radioactive GABA readily accumulates in the median eminence [44]. Thus, the action of GABA in this area might account for the central effect when administered

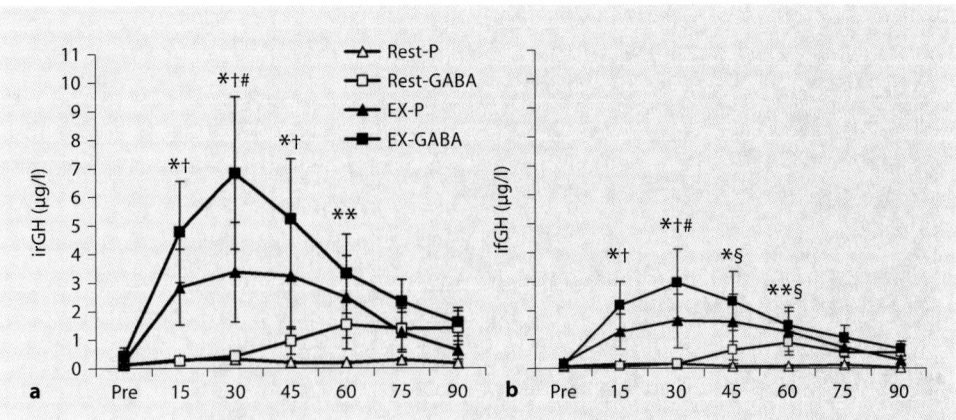

Fig. 4. Immunoreactive growth hormone (irGH) (**a**) and immunofunctional GH (ifGH) (**b**) time-point concentrations for the rest-placebo (P), rest-GABA, exercise (EX)-P, and EX-GABA conditions. * EX-GABA different from rest-P and rest-GABA ($p < 0.01$); [†] EX-P different from rest-P and rest-GABA ($p < 0.05$); [#] EX-GABA different from EX-P ($p < 0.05$); ** EX-GABA different from rest-P ($p < 0.01$); [§] EX-P different from rest-P ($p < 0.05$) [from 26].

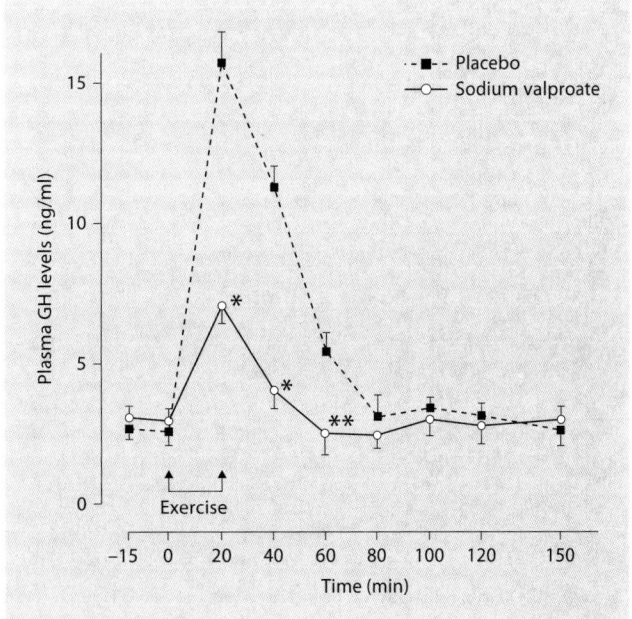

Fig. 5. GH response to cycling exercise following oral administration of sodium valproate or placebo [from 28].

peripherally. Likewise, the pituitary gland sits outside the blood-brain barrier, allowing for direct action by peripheral GABA. Researchers investigating the effects of oral GABA on GH have failed to assess the GABA concentration of the plasma. Thus, it is possible that oral GABA undergoes liver-induced biotransformation to other amino acids which may also stimulate GH secretion [5]. Regardless of the mechanisms involved, it seems that GABA ingestion results in increased GH concentrations, both at rest and after resistance exercise.

Conclusion

The precise nature of GABA's effect on GH secretion as well as its mechanism of action remains to be clarified. As GABA ingestion apparently stimulates GH release, it is certainly possible that GABA induces lipolytic effects and skeletal muscle protein accretion, via mechanisms directly and/or indirectly related to GH release [1]. However, the anabolic and lipolytic values of the relatively small GABA-induced GH responses remains unclear. It is plausible that GABA supplementation stimulates GHRH or ghrelin secretion and/or suppresses SS release, as previously demonstrated in animal models. To date, no studies have investigated the ergogenic value of GABA ingestion. Given the acute GH responses to GABA ingestion, studies designed to determine the effects of longitudinal GABA supplementation on anthropometric and performance measures are warranted.

References

1 Godfrey RJ, Madgwick Z, Whyte GP: The exercise-induced growth hormone response in athletes. Sports Med 2003;33:599–613.
2 Baumann GP: Growth hormone isoforms. Growth Horm IGF Res 2009;19:333–340.
3 Veldhuis JD, Bowers CY: Human GH pulsatility: an ensemble property regulated by age and gender. J Endocrinol Invest 2003;26:799–813.
4 Wideman L, Weltman JY, Hartman ML, Veldhuis JD, Weltman A: Growth hormone release during acute and chronic aerobic and resistance exercise: recent findings. Sports Med 2002;32:987–1004.
5 Chromiak JA, Antonio J: Use of amino acids as growth hormone-releasing agents by athletes. Nutrition 2002;18:657–661.
6 Tannenbaum GS, Epelbaum J, Bowers CY: Interrelationship between the novel peptide ghrelin and somatostatin/growth hormone-releasing hormone in regulation of pulsatile growth hormone secretion. Endocrinology 2003;144:967–974.l
7 Burt DR, Kamatchi GL: GABA$_A$ receptor subtypes: from pharmacology to molecular biology. FASEB J 1991;5:2916–2923.
8 Ong J, Kerr DI: Recent advances in GABA$_B$ receptors: from pharmacology to molecular biology. Acta Pharmacol Sin 2000;21:111–123.
9 Gamel-Didelon K, Kunz L, Fohr KJ, Gratzl M, Mayerhofer A: Molecular and physiological evidence for functional γ-aminobutyric acid (GABA$_C$) receptors in growth hormone-secreting cells. J Biol Chem 2003;278:20192–20195.
10 Schousboe A, Westergaard N, Waagepetersen HS, Larsson OM, Bakken IJ, Sonnewald U: Trafficking between glia and neurons of TCA cycle intermediates and related metabolites. Glia 1997;21:99–105.
11 Petroff OA: GABA and glutamate in the human brain. Neuroscientist 2002;8:562–573.
12 Okada Y, Nitsch-Hassler C, Kim JS, Bak IJ, Hassler R: Role of γ-aminobutyric acid (GABA) in the extrapyramidal motor system. 1. Regional distribution of GABA in rabbit, rat, guinea pig and baboon CNS. Exp Brain Res 1971;13:514–518.
13 Acs Z, Lonart G, Makara GB: Role of hypothalamic factors (growth hormone-releasing hormone and γ-aminobutyric acid) in the regulation of growth hormone secretion in the neonatal and adult rat. Neuroendocrinology 1990;52:156–160.
14 Acs Z, Szabo B, Kapocs G, Makara GB: γ-Aminobutyric acid stimulates pituitary growth hormone secretion in the neonatal rat. A superfusion study. Endocrinology 1987;120:1790–1798.

15 Acs Z, Zsom L, Makara GB: Possible mediation of GABA-induced growth hormone secretion by increased calcium flux in neonatal pituitaries. Life Sci 1992;50:217–279.

16 Anderson RA, Mitchell R: Effects of γ-aminobutyric acid receptor agonists on the secretion of growth hormone, luteinizing hormone, adrenocorticotrophic hormone and thyroid-stimulating hormone from the rat pituitary gland in vitro. J Endocrinol 1986;108:1–8.

17 Baes M, Vale WW: Growth hormone-releasing factor secretion from fetal hypothalamic cell cultures is modulated by forskolin, phorbol esters, and muscimol. Endocrinology 1989;124:104–110.

18 Cavagnini F, Benetti G, Invitti C, et al: Effect of γ-aminobutyric acid on growth hormone and prolactin secretion in man: influence of pimozide and domperidone. J Clin Endocrinol Metab 1980;51:789–792.

19 Cavagnini F, Invitti C, Pinto M, et al: Effect of acute and repeated administration of γ-aminobutyric acid on growth hormone and prolactin secretion in man. Acta Endocrinol (Copenh) 1980;93:149–154.

20 Cavagnini F, Invitti C, Di Landro A, Tenconi L, Maraschini C, Girotti G: Effects of a γ-aminobutyric acid derivative, baclofen, on growth hormone and prolactin secretion in man. J Clin Endocrinol Metab 1977;45:579–584.

21 Coiro V, Volpi R, Maffei ML, et al: Opioid modulation of the γ-aminobutyric acid-controlled inhibition of exercise-stimulated growth hormone and prolactin secretion in normal men. Eur J Endocrinol 1994;131:50–55.

22 Fiók J, Acs Z, Makara GB, Erdö SL: Site of γ-aminobutyric acid-mediated inhibition of growth hormone secretion in the rat. Neuroendocrinology 1984;39:510–516.

23 Gudelsky GA, Apud JA, Masotto C, Locatelli V, Cocchi D, Racagni G, Müller EE: Ethanolamine-O-sulfate enhances γ-aminobutyric acid secretion into hypophysial portal blood and lowers serum prolactin concentrations. Neuroendocrinology 1983;37:397–399.

24 Koulu M, Lammintausta R, Dahlström S: Effects of some γ-aminobutyric acid (GABA)ergic drugs on the dopaminergic control of human growth hormone secretion. J Clin Endocrinol Metab 1980;51:124–129.

25 Monteleone P, Maj M, Iovino M, Steardo L: Evidence for a sex difference in the basal growth hormone response to GABAergic stimulation in humans. Acta Endocrinol (Copenh) 1988;119:353–357.

26 Powers ME, Yarrow JF, McCoy SC, Borst SE: Growth hormone isoform responses to GABA ingestion at rest and after exercise. Med Sci Sports Exerc 2008;40:104–110.

27 Spencer GS, Berry CJ, Bass JJ: Neuroendocrine regulation of growth hormone secretion in sheep. VII. Effects of GABA. Regul Pept 1994;52:181–186.

28 Steardo L, Iovino M, Monteleone P, Agrusta M, Orio F: Pharmacological evidence for a dual GABAergic regulation of growth hormone release in humans. Life Sci 1986;39:979–985.

29 Takahara J, Yunoki S, Hosogi H, Yakushiji W, Kageyama J, Ofuji T: Concomitant increases in serum growth hormone and hypothalamic somatostatin in rats after injection of γ-aminobutyric acid, aminooxyacetic acid, or γ-hydroxybutyric acid. Endocrinology 1980;106:343–347.

30 Vijayan E, McCann SM: Effects of intraventricular injection of γ-aminobutyric acid on plasma growth hormone and thyrotropin in conscious ovariectomized rats. Endocrinology 1978;103:1888–1893.

31 Willoughby JO, Jervois PM, Menadue MF, Blessing WW: Activation of GABA receptors in the hypothalamus stimulates secretion of growth hormone and prolactin. Brain Res 1986;374:119–125.

32 Sakaue M, Saito N, Taniguchi H, Baba S, Tanaka C: Immunohistochemical localization of γ-aminobutyric acid in the rat pituitary gland and related hypothalamic regions. Brain Res 1988;446:343–353.

33 Willoughby JO, Beroukas D, Blessing WW: Ultrastructural evidence for γ-aminobutyric acid-immunoreactive synapses on somatostatin-immunoreactive perikarya in the periventricular anterior hypothalamus. Neuroendocrinology 1987;46:268–272.

34 Meister B, Hokfelt T: Peptide- and transmitter-containing neurons in the mediobasal hypothalamus and their relation to GABAergic systems: possible roles in control of prolactin and growth hormone secretion. Synapse 1988;2:585–605.

35 Racagni G, Apud JA, Cocchi D, Locatelli V, Muller EE: GABAergic control of anterior pituitary hormone secretion. Life Sci 1982;31:823–838.

36 Tappaz ML, Wassef M, Oertel WH, Paut L, Pujol JF: Light- and electron-microscopic immunocytochemistry of glutamic acid decarboxylase in the basal hypothalamus: morphological evidence for neuroendocrine γ-aminobutyrate. Neuroscience 1983;9:271–287.

37 Zemkova HW, Bjelobaba I, Tomic M, Zemkova H, Stojilkovic SS: Molecular, pharmacological and functional properties of $GABA_A$ receptors in anterior pituitary cells. J Physiol 2008;586:3097–3111.

38 Mitchell R, Grieve G, Dow R, Fink G: Endogenous GABA receptor ligands in hypophysial portal blood. Neuroendocrinology 1983;37:169–176.

39 Fiszer de Plazas S, Becú D, Mitridate de Novara A, Libertun C: Gamma-aminobutyric acid receptors in anterior pituitary and brain areas after median eminence lesions. Endocrinology 1982;111:1974–1978.

40 Schimchowitsch S, Vuillez P, Tappaz ML, Klein MJ, Stoeckel ME: Systematic presence of GABA-immunoreactivity in the tuberoinfundibular and tuberohypophyseal dopaminergic axonal systems: an ultrastructural immunogold study on several mammals. Exp Brain Res 1991;83:575–586.

41 Mayerhofer A, Höhne-Zell B, Gamel-Didelon K, Jung H, Redecker P, Grube D, Urbanski HF, Gasnier B, Fritschy JM, Gratzl M: Gamma-aminobutyric acid: a para- and/or autocrine hormone in the pituitary. FASEB J 2001;15:1089–1091.

42 Rage F, Benyassi A, Arancibia S, Tapia-Arancibia L: Gamma-aminobutyric acid-glutamate interaction in the control of somatostatin release from hypothalamic neurons in primary culture: in vivo corroboration. Endocrinology 1992;130:1056–1062.

43 Vijayan E, McCann SM: Blockade of dopamine receptors with pimozide and pituitary hormone release in response to intraventricular injection of γ-aminobutyric acid in conscious ovariectomized rats. Brain Res 1979;162:69–76.

44 Kuroda E, Watanabe M, Tamayama T, Shimada M: Autoradiographic distribution of radioactivity from ^{14}C-GABA in the mouse. Microsc Res Tech 2000;48:116–126.

Michael Powers, PhD, ATC, EMT, CSCS
Marist College, 3399 North Road
Poughkeepsie, NY 12601 (USA)
Tel. +1 845 575 3912
E-Mail Michael.Powers@marist.edu

Lamprecht M (ed): Acute Topics in Sport Nutrition.
Med Sport Sci. Basel, Karger, 2013, vol 59, pp 47–56

Exercise, Intestinal Barrier Dysfunction and Probiotic Supplementation

Manfred Lamprecht[a,b] · Anita Frauwallner[c,d]

[a]Institute of Physiological Chemistry, Center for Physiological Medicine, Medical University of Graz, and
[b]Institute of Nutrient Research and Sport Nutrition, Graz, and [c]Austrian Society of Probiotic Medicine, Vienna,
and [d]Institut Allergosan, Forschungs- und Vertriebs GmbH, Graz, Austria

Abstract

Athletes exposed to high-intensity exercise show an increased occurrence of gastrointestinal (GI) symptoms like cramps, diarrhea, bloating, nausea, and bleeding. These problems have been associated with alterations in intestinal permeability and decreased gut barrier function. The increased GI permeability, a so-called 'leaky gut', also leads to endotoxemia, and results in increased susceptibility to infectious and autoimmune diseases, due to absorption of pathogens/toxins into tissue and the bloodstream. Key components that determine intestinal barrier function and GI permeability are tight junctions, protein structures located in the paracellular channels between epithelial cells of the intestinal wall. The integrity of tight junctions depends on sophisticated interactions between the gut residents and their expressed substances, the intestinal epithelial cell metabolism and the activities of the gut-associated lymphoid tissue. Probiotic supplements are an upcoming group of nutraceuticals that could offer positive effects on athlete's gut and entire health. Some results demonstrate promising benefits for probiotic use on the athlete's immune system. There is also evidence that probiotic supplementation can beneficially influence intestinal barrier integrity in acute diseases. With regard to exercise-induced GI permeability problems, there is still a lack of studies with appropriate data and a gap to understand the underlying mechanisms to support such health beneficial statements implicitly. This article refers (i) to exercise-induced intestinal barrier dysfunction, (ii) provides suggestions to estimate increased gut barrier permeability in athletes, and (iii) discusses the potential of probiotic supplementation to counteract an exercise-induced leaky gut.

In performance sports there is a high prevalence of gastrointestinal (GI) complaints, especially among endurance athletes like runners and triathletes [1]. These problems are attributed to changed blood flow that is shunted from the viscera to skeletal muscle or the heart [2]. Symptoms described are nausea, stomach and intestinal cramps, vomiting and diarrhea. An increased permeability of the GI epithelial wall precedes and accompanies these symptoms. This so-called 'leaky gut' leads to endotoxemia, and results in increased susceptibility to infectious and auto-immune diseases, due to absorption of pathogens/toxins into tissue and the blood-stream [3–5].

Probiotic bacteria are described as live microorganisms that beneficially modulate microflora and health of the host [6]. In the last few years they have become increasingly popular as nutritional supplements, especially to achieve reduction of GI complaints and common infectious illnesses. In sports and exercise, there is some evidence for probiotics' potential to reduce the incidence and severity of respiratory tract infections [7, 8] and to shorten the duration of GI symptoms in trained athletes [9]. Other studies report attenuation of exercise-induced increase in pro- and anti-inflammatory cytokines after 11 weeks [10] and increased plasma antioxidant levels after 4 weeks of probiotic supplementation [11].

The effects of probiotics on intestinal barrier integrity are not sufficiently elucidated and even less in the framework of sports and exercise. Our research using PubMed did not reveal any study reporting the effects of probiotic supplementation on exercise-induced gut barrier dysfunction, although such nutritional solutions could be of reasonable practical relevance for athletes.

This lack in sport nutrition research encouraged us to write this article dealing with exercise-induced intestinal barrier dysfunction and discussing the potential of probiotic supplementation to counteract the associated problems.

Intestinal Barrier

The intestinal mucosa is the largest boundary layer between the surrounding world and the internal human milieu representing a surface area similar to the size of a soccer field [12]. The ability to control and balance the invasion of harmful content from the intestinal lumen into the internal organism is called intestinal barrier function. This barrier can roughly be divided into three parts where (i) luminal bacteria, (ii) epithelial cells, and (iii) the gut-associated lymphoid tissue (GALT) work in concert to prevent potential pathogens from translocation from the lumen into the internal milieu [13] (fig. 1).

(i) Most bacteria within the GI tract are killed by bactericidal salivary lysozyme, gastric acid and pepsin, bile salts and Paneth cell-derived defensins [14]. If bacteria survive this security system they are combated by the commensal microbiota to inhibit their colonization. Commensal bacteria produce antimicrobial substances to kill competitor strains, they induce pH modification of the luminal content and act competitively to get nutrients required for bacterial growth [15].

(ii) The epithelial layer covers the inner surface of the intestine. These epithelial cells are connected to each other by junctional complexes that consist of tight junctions, adherens junctions and desmosomes. The tight junctions are the most important physical barrier protein structures that allow selective passage of

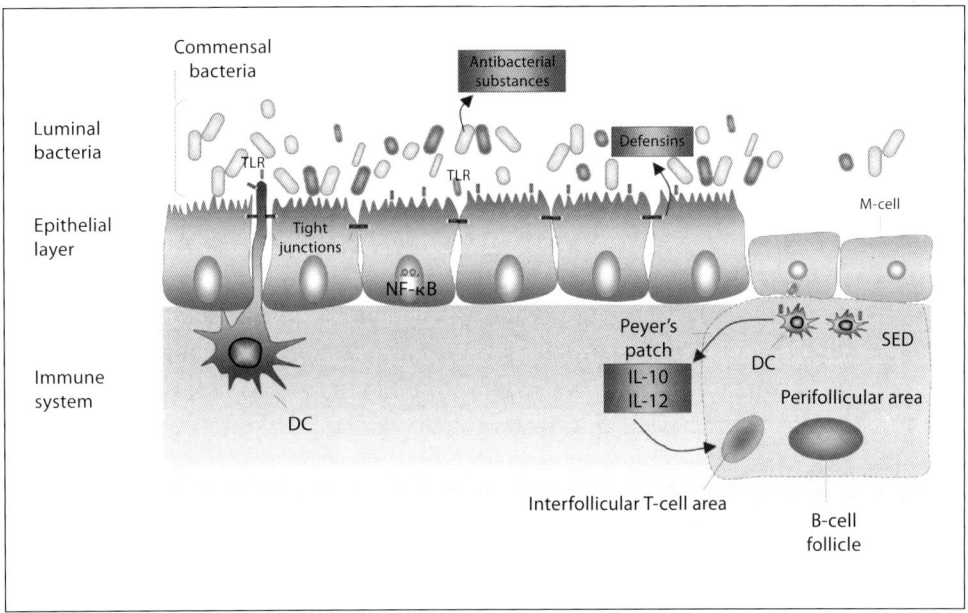

Fig. 1. Three levels of barrier function (from F. Lutgendorff: Defending the barrier [13]): (1) Luminal commensal bacteria prevent colonization of potential pathogens by producing antibacterial substances and by forming competition for nutrients required for bacterial growth of pathogenic microorganisms. (2) Epithelial cell form the most obvious physical barrier and are interconnected by means of tight junctions. Recognition of pathogens and commensal bacteria is mediated through differential activation of Toll-like receptors (TLRs). (3) Upon recognition of potential pathogens by TLRs, epithelial cells can respond with an innate reaction to activate nuclear factor NF-κB. Dendritic cells (DCs) are key players in the adaptive immune response and reside primarily within the subepithelial dome (SED) of Peyer's patches, which is covered with specialized M-cells. Once potential pathogens are recognized by DCs, these respond with the production of interleukin-12, skewing the T-cell response towards a proinflammatory Th1-type response.

ions and small molecules and they form a tight barrier to protein-sized molecules and bacteria [16]. Beside the physical barrier function, the epithelial layer is also responsible for the communication between the luminal contents and the basolateral located gut-associated lymphoid tissue. Epithelial cells are equipped with Toll-like receptors, which are capable of recognizing both commensal and pathogenic bacteria and their toxic products such as lipopolysaccharides. When detected by Toll-like receptors, potential pathogens induce proinflammatory mechanisms within epithelial cells, leading to an immune response directed against these intruders [17].

(iii) The GALT is the largest immune organ in the body and most of it is organized in structures named Peyer's patches. The covering epithelium of the Peyer's patches contains M-cells, which possess the ability to take up antigens and bacteria from the intestinal lumen through endocytosis and deliver these to antigen-presenting cells and lymphocytes located in the basolateral side of the Peyer's patches [18]. Below the epithelial layer, dendritic cells are mainly responsible for the decision to either ignore or to respond to invading antigens and bacteria [13]. They orchestrate immune responses by instructing T-cells to express pro- or anti-inflammatory cytokines like IL-12 or IL-10.

Exercise and Disturbed Intestinal Barrier Function

Strenuous exercise can lead to functional disturbances in the complex barrier system of the intestinal wall. The main causative base that triggers decreased barrier function during exercise might be a changed blood flow, which is shunted from the viscera to skeletal muscle and the heart [2]. The higher the intensity and the longer the duration of the conducted exercise model, the more distinctly this changed blood flow occurs in the gut, heart and skeletal muscle regions. A mesenteric undersupply with blood, oxygen, nutrients and the insufficient removal of metabolites leads to a metabolic disturbance in the intestinal barrier area including hypoperfusion/ischemia-induced pH decrease which in turn results in oxidative stress. Also, the increased production of catecholamines during intensive exercise and their subsequent autoxidation may contribute to disturbing the mesenteric redox balance. The overwhelming production of reactive oxygen and nitrogen species (RONS), including hydrogen peroxide, results in oxidized and damaged biomolecules, changed cell communication pathways and inflammatory processes induced by cytokines from GALT-based immune cells. Subsequently, tight junction proteins and epithelial cell membranes are some of the affected structures that change and/or lose their integrity [19, 20]. In addition, there might be some other exercise-induced disturbances in the mesenteric area like water availability (dehydration), changed osmolality and gut motility that can contribute to a decreased intestinal barrier integrity [21].

Decreased tight junction competency leads to a 'leak' in the paracellular absorption route that enables pathogens/toxins to pass from the intestinal milieu, challenging the immune system to produce or enhance an immunological response, accompanied by inflammatory processes and oxidative stress [22–24]. This event is called endotoxemia and has important meanings for performance athletes because it increases susceptibility to infectious and autoimmune diseases [3–5]. We postulate that a disturbed intestinal barrier function in athletes represents one of the main triggers of immune reactions with the often reported consequences of increased susceptibility to infectious diseases. If carers observe that their athletes show a conspicuous frequency of common colds and allergic reactions they should consider an estimation of intestinal mucosal barrier function and appropriate dietary interventions to recover from a detrimentally increased gut permeability.

Surrogate Markers of Practical Relevance to Estimate Intestinal Barrier Function

Several methods and markers are described to estimate intestinal barrier integrity like for example determination of the transepithelial electrical resistance (TEER) of epithelial monolayers or measurement of tight junction molecules like claudins, occludin or zonula occludens proteins [13, 19]. However, from our subjective perspective, the currently most appropriate surrogate marker that represents and modulates intestinal barrier function/tight junction integrity in humans, which is also easy to measure in blood and feces, seems to be zonulin. Zonulin – a protein of the haptoglobin family – is described as a physiological modulator of intercellular tight junctions and increased concentrations are related to changes in tight junction competency and increased GI permeability [19, 25]. Beside liver cells, intestinal cells can synthesize zonulin and the most powerful triggers to activate the zonulin system are postulated to be dietary proteins like gliadins (e.g. in gluten-rich cereals) or enteric bacteria [19, 25, 26]. Zonulin communicates with the epidermal growth factor receptor and several G-protein-coupled receptors like proteinase activating receptor 2 on the intestinal cell walls. Its signal transduction pathway causes tight junctions disassembly. In a recent review, Fasano [25] postulates that the zonulin-dependent regulation of tight junctions is involved in regulation of the movement of fluids and ions, macromolecules and leukocytes between the bloodstream and the intestinal lumen and vice versa. In addition, he suggests that the intestinal zonulin pathway is needed to avoid microorganism colonization of the proximal intestine.

It is important to note at this point that zonulin is also a suitable marker to estimate exercise-induced intestinal barrier disturbances as we were able to observe increased zonulin concentrations in feces of athletes recently [27]. These are findings of high practicability for carers of athletes because there are non-sophisticated analytical methods to measure zonulin (ELISA), and some generally accessible laboratories in Western Europe and the USA already provide services routinely to estimate zonulin concentrations from feces or serum. Hence, we postulate in line with other researchers, that zonulin can be regarded as one of the most valid surrogate markers to estimate intestinal barrier function in humans [28–30].

Plasma lipopolysaccharide (LPS) is another marker to estimate intestinal barrier function, although it is a more general one as many cells release these molecules. Nevertheless, high LPS concentrations correlate with increased intestinal permeability [31]. LPSs are proinflammatory agents that activate the transcriptional factor NF-κB to induce the production of inflammatory mediators such as proinflammatory cytokines, chemokines, nitric oxide and RONS [32]. A colonization of the intestine with unfavorable Gram-negative microbiota leads to increased plasma LPS levels, a so-called 'metabolic endotoxemia', whereas the quantity of *Bacteroidetes* species decreases and that of *Firmicutes* bacteria increases [33]. The Gram-negative-derived LPSs are transported from the intestinal lumen towards the target tissue by a mechanism facilitated by chylomicrons that are synthesized from the intestinal epithelial cells. The LPSs then bind to Toll-like-4 receptors on macrophages and trigger secretion of proinflammatory cytokines [31]. Increased intestinal permeability goes hand in hand with translocation of LPSs from the intestine to the blood and elevated LPS concentrations can be determined via ELISA methods in plasma.

Potential of Probiotic Supplementation to Combat an Exercise-Induced Leaky Gut

Athletes exposed to high-intensity exercise show an increased occurrence of GI symptoms like cramps, diarrhea, bloating, nausea, and bleeding [34, 35]. These symptoms have been associated with alterations in intestinal permeability and decreased barrier function [36, 37] as well as with inflammation and oxidative stress [23, 25]. Some researchers also postulate that the Peyer's patches of the GI tract are linked to diverse effector sites of other mucosal surfaces such as the upper respiratory tract and urogenital tract [3]. Such proposed liaisons within different mucosal layers suggest that general immunity depends partially from the immunologic situation in the intestinal wall, and common perception exists that some athletes in rowing, cycling, running, swimming, cross-country skiing, biathlon, triathlon but also team sports (soccer, ice hockey, basketball) may suffer from an increased incidence of upper respiratory tract illness during heavy training and competition.

From the described observations, together with the fact that the intestinal barrier function is affected by microbiota's metabolism, arises the hypothesis that appropriate probiotic supplementation could have potential to counteract exercise-induced intestinal barrier dysfunction. Although we found a few exercise-associated studies on probiotic supplementation and its influence on infectious diseases and GI symptoms [7–10], till now we could not find a study that investigated the influence of probiotic supplementation on athletes' intestinal barrier function directly.

However, right now we finalized a study with trained men that investigated the effects of 14 weeks' multi-species probiotic supplementation (Ecologic®Performance/OMNi-BiOTiC®Power) on zonulin concentrations in feces [27]. Our data showed significantly decreased stool zonulin concentrations after 14 weeks of probiotic supplementation. The observed reduction is all the more remarkable as the mean concentrations were slightly above normal at baseline (reference range <30 ng•ml^{-1}) and dropped to normal after 14 weeks with probiotics. These data indicate that the trained men of our investigated cohort suffered already a mild increase in intestinal permeability at baseline. This was probably affected by their chronic exercise training. We hypothesize that the effects of the supplemented probiotics surpassed bacteria that activate the zonulin system and could settle in the deep intestine and/or even replaced zonulin-stimulating bacteria. This observation on zonulin decrease after probiotic supplementation is of practical relevance for athletes, especially under the perspective that an improved intestinal barrier reduces the athlete's susceptibility to endotoxemia and associated cytokine production [38].

To round down this topic on probiotic supplementation and exercise-induced leaky gut, it is necessary to discuss the appropriate bacterial strains, the best matrix in which probiotics are provided, and concentrations/dosages to increase the probability for efficacy. In our study [27] we used six strains: *Bifidobacterium bifidum* W23, *Bifidobacterium lactis* W51, *Enterococcus faecium* W54, *Lactobacillus acidophilus* W22, *Lactobacillus brevis* W63, and *Lactococcus lactis* W58. This mixture of strains demonstrated appropriate efficacy on zonulin reduction but there are also scientific data on other effective strains, e.g. *Bifidobacterium infantis* Y1 [39], several strains from *Lactobacillus plantarum* [40–42], *Bacteroides thetaiotaomicron* ATCC29184 [43], *Escherichia coli* Nissle 1917 [44], *Bifidobacterium longum* SP 07/3 and *Lactobacillus rhamnosus* GG [45] revealed beneficial impacts on intestinal barrier function.

Concerning the dosage of probiotic supplementation, we applied a daily concentration of 10^{10} CFU (colony-forming units), which we regard as an appropriate minimum dose. The powder matrix we used consisted of cornstarch, maltodextin and minerals. It is commonly accepted that the matrix is essential to modulate the metabolism of the strains and hence to ensure stability, gastric survival rate and probability to settle and to act bioeffective in the deep intestine. We hypothesize that the composition of the matrix of a probiotic product might be as important as the contained strain(s) itself to guarantee bioefficacy. Despite this important role of the probiotics' matrix, it is interesting that there is little scientific literature that refers to that issue.

Summary, Conclusion and Perspective

Intestinal barrier dysfunction occurs in high-performance exercise. A leaky gut leads to several GI complaints and increases susceptibility to allergies and infectious diseases. The base causality for a gut barrier dysfunction seems to trace to a considerably reduced blood flow during exercise. The blood undersupply induces several metabolic disturbances in the mesenteric region like increased pH value, oxidative stress and inflammation. The metabolic changes favor the communication between bacteria and intestinal epithelium which reacts with increased zonulin expression. Zonulin modulates tight junction disassembly and consequently increased intestinal wall permeability. Athletes who struggle with a high incidence of infectious diseases and allergic reactions should consider monitoring intestinal barrier function via regular zonulin and LPS determination from feces or serum/plasma. Several strains and multi-species probiotic supplementation with a minimum daily dose of 10^{10} CFU showed potential to benefit intestinal barrier function and reduced zonulin concentrations in the stool of trained men.

In conclusion, the few existing data support the hypothesis that an adequate probiotic supplementation can improve intestinal barrier integrity in sporty people. Subsequent sport studies that focus on leaky gut and associated consequences like endotoxemia, inflammation, athlete's susceptibility to infections and allergies, will reveal more clarity. Such future data from research, and the development of appropriate products and modes of application, will provide high practical relevance for athletes, doctors, physiotherapists and other carers.

Disclosure Statement

M. Lamprecht has been awarded competitive research grants from Winclove BV, Amsterdam, The Netherlands, and the Institute Allergosan, Forschungs- und Vertriebs GmbH, Graz, Austria, to study Ecologic®Performance/OMNi-BiOTiC®Power (produced by Winclove BV).

A. Frauwallner is the CEO of Institut Allergosan, Forschungs- und VertriebsGmbH, Graz, Austria, a distributor of OMNi-BiOTiC®Power.

References

1 Rehrer NJ, Brouns F, Beckers EJ, Frey WO, Villiger B, Riddoch CJ, Menheere PP, Saris WH: Physiological changes and gastrointestinal symptoms as a result of ultra-endurance running. Eur J Appl Physiol Occup Physiol 1992;64:1–8.

2 Qarnar MI, Read AE: Effects of exercise on mesenteric blood flow in man. Gut 1987;28:583–587.

3 West NP, Pyne DB, Peake JM, Cripps AW: Probiotics, immunity and exercise: a review. Exerc Immunol Rev 2009;15:107–126.

4 Fasano A: Leaky gut and autoimmune diseases. Clinic Rev Allerg Immunol 2012;42:71–78.

5 DeOliveira EP, Burini RC: Food-dependent, exercise-induced gastrointestinal distress. J Int Soc Sports Nutr 2011;8:12.

6 Salminen S, Bouley D, Bourron-Ruault MC, Cummings JH, Franck A, Gibson GR, Isolauri E, Moreau MC, Roberfroid M, Rowland I: Functional food science and gastrointestinal physiology and function. Br J Nutr 1998;80(suppl):S147–S171.

7 Gleeson M, Bishop NC, Oliveira M, Tauler P: Daily probiotic's (Lactobacillus casei Shirota) reduction of infection incidence in athletes. Int J Sport Nutr Exerc Metab 2011;21:55–64.

8 Cox AJ, Pyne DB, Saunders PU, Fricker PA: Oral administration of the probiotic Lactobacillus fermentum VRI-003 and mucosal immunity in endurance athletes. Br J Sports Med 2010;44:222–226.

9 Kekkonen RA, Vasankari TJ, Vuorimaa T, Haahtela T, Julkunen I, Korpela R: The effect of probiotics on respiratory infections and gastrointestinal symptoms during training in marathon runners. Int J Sport Nutr Exerc Metab 2007;17:352–363.

10 West NP, Pyne DB, Cripps AW, Hopkins WG, Eskesen DC, Jairath A, Christophersen CT, Conlon MA, Fricker PA: Lactobacillus fermentum (PCC®) supplementation and gastrointestinal and respiratory-tract illness symptoms: a randomised control trial in athletes. Nutr J 2011;10:30.

11 Martarelli D, Verdenelli MC, Scuri S, Cocchioni M, Silvi S, Cecchini C, Pompei P: Effect of a probiotic intake on oxidant and antioxidant parameters in plasma of athletes during intense exercise training. Curr Microbiol 2011;62:1689–1696.

12 Kalliomaki M, Isolauri E: Role of intestinal flora in the development of allergy. Curr Opin Allergy Clin Immunol 2003;1:15–20.

13 Lutgendorff F: Defending the barrier; effects of probiotics on endogenous defense mechanisms; thesis, Utrecht. Enschede, Gilderprint Drukkerijen, 2009. ISBN/EAN 978–94–901–2285–0.

14 Shanahan F: The host-microbe interface within the gut. Best Pract Res Clin Gastroenterol 2002;6: 915–931.

15 Hooper LV, Xu J, Falk PG, Midtvedt T, Gordon JI: A molecular sensor that allows a gut commensal to control is nutrient foundation in a competitive ecosystem. Proc Natl Acad Sci USA 1999;17:9833–9838.

16 Van Itallie CM, Anderson JM: Claudins and epithelial paracellular transport. Annu Rev Physiol 2006; 68:403–429.

17 Takeda K, Kaisho T, Akira S: Toll-like receptors. Annu Rev Immunol 2003;21:335–376.

18 Gebert A: The role of M cells in the protection of mucosal membranes. Histochem Cell Biol 1997;6: 455–470.

19 Ulluwishewa D, Anderson RC, McNabb WC, Moughan PJ, Wells JM, Roy NC: Regulation of tight junction permeability by intestinal bacteria and dietary components. J Nutr 2011;141:769–776.

20 Basuroy S, Seth A, Elias B, Naren AP, Rao R: MAPK interacts with occludin and mediates EGF-induced prevention of tight junction disruption by hydrogen peroxide. Biochem J 2006;393:69–77.

21 Alverdy JC, Laughlin RS, Wu L: Influence of the critically ill state on host-pathogen interactions within the intestine: gut-derived sepsis redefined. Crit Care Med 2003;2:598–607.

22 Fasano A: Pathological and therapeutical implications of macro-molecule passage through the tight junction; in Tight Junctions. Boca Raton, CRC Press, 2001, pp 697–722.

23 Groschowitz KR, Hogan SP: Intestinal barrier function: molecular regulation and disease pathogenesis. J Allergy Clin Immunol 2009;124:3–20.

24 Sonier B, Patrick C, Ajjikuttira P, Scott FW: Intestinal immune regulation as a potential diet-modifiable feature of gut inflammation and autoimmunity. Int Rev Immunol 2009;28:414–445.

25 Fasano A: Zonulin and its regulation of intestinal barrier function: the biological door to inflammation, autoimmunity, and cancer. Physiol Rev 2011; 91:151–175.

26 El Asmar R, Panigrahi P, Bamford P, Berti I, Not R, Coppa GV, Catassi C, Fasano A: Host-dependent activation of the zonulin system is involved in the impairment of the gut barrier function following bacterial colonization. Gastroenterology 2002;123: 1607–1615.

27 Lamprecht M, Bogner S, Schippinger G, Steinbauer K, Fankhauser F, Hallstroem S, Schuetz B, Greilberger JF: Probiotic supplementation affects markers of intestinal barrier, oxidation, and inflammation in trainer men; a randomized, double-blinded, placebo-controlled trial. J Int Soc Sports Nutr 2012 (submitted).

28 Fasano A: Intestinal zonulin: open sesame! Gut 2001;49:159–162.

29 Sapone A, de Magistris L, Pietzak M, Clemente MG, Tripathi A, Cucca F, Lampis R, Kryszak D, Carteni M, Generoso M, Iafusco D, Prisco F, Laghi F, Riegler G, Carratu R, Counts D, Fasano A: Zonulin upregulation is associated with increased gut permeability in subjects with type 1 diabetes and their relatives. Diabetes 2006;55:1443–1449.

30 Wang W, Uzzau S, Goldblum SE, Fasano A: Human zonulin, a potential modulator of intestinal tight junctions. J Cell Sci 2000;113:4435–4440.

31 DeKort S, Keszthelyi D, Masclee AAM: Leaky gut and diabetes mellitus: what is the link? Obes Rev 2011;12:449–458.

32 Wheeler MD: Endotoxin and Kupffer cell activation in alcoholic liver disease. Alcohol Res Health 2003;27:300–306.

33 Ley RE, Backhed F, Turnbaugh P, Lozupone CA, Knight RD, Gordon JI: Obesity alters gut microbial ecology. Proc Natl Acad Sci USA 2005;102:11070–11075.

34 Baska RS, Moses FM, Graeber G, Kearney G: Gastrointestinal bleeding during an ultramarathon. Dig Dis Sci 1990;35:276–279.

35 Eichner ER: Gastrointestinal bleeding in athletes. Phys Sportsmed 1989;17:128–140.

36 Oktedalen O, Lunde OC, Opstad PK, Aabakken L, Kvernebo K: Changes in the gastrointestinal mucosa after long-distance running. Scand J Gastroenterol 1992;27:270–274.

37 Pals KL, Chang RT, Ryan AJ, Gisolfi CV: Effect of running intensity on intestinal permeability. J Appl Physiol 1997;82:571–576.

38 Jeukendrup AE, Vet-Joop K, Sturk A, Stegen JHJC, Senden J, Saris WHM, Wagenmakers AJM: Relationship between gastrointestinal complaints and endotoxaemia, cytokine release and the acute-phase reaction during and after a long-distance triathlon in highly trained men. Clin Sci 2000;98: 47–55.

39 Ewaschuk JB, Diaz H, Meddings L, Diederichs B, Dmytrash A, Backer J, Looijer-van Langen M, Madsen KL: Secreted bioactive factors from *Bifidobacterium infantis* enhance epithelial cell barrier function. Am J Physiol Gastrointest Liver Physiol 2008;295:G1025–1034.

40 Qin H, Zhang Z, Hang X, Jiang YL: *L. plantarum* prevents enteroinvasive *Escherichia coli*-induced tight junction proteins changes in intestinal epithelial cells. BMC Microbiol 2009;9:63.

41 Anderson RC, Cookson AL, McNabb WC, Kelly WJ, Roy NC: Lactobacillus plantarum DSM 2648 is a potential probiotic that enhances intestinal barrier function. FEMS Microbiol Lett 2010;309: 184–192.

42 Karczewski J, Troost FJ, Konings I, Dekker J, Kleerebezem M, Brummer RJM, Wells JM: Regulation of human epithelial tight junction proteins by *Lactobacillus plantarum* in vivo and protective effects on the epithelial barrier. Am J Physiol Gastrointest Liver Physiol 2010;298:G851–G859.

43 Resta-Lenert S, Barrett KE: Probiotics and commensals reverse TNF-α- and IFN-γ-induced dysfunction in human intestinal epithelial cells. Gastroenterology 2006;130:731–746.

44 Ukena SN, Singh A, Dringenberg U, Engelhardt R, Seidler U, Hansen W, Bleich A, Bruder D, Franzke A, Rogler G, et al: Probiotic *Escherichia coli* Nissle 1917 inhibits leaky gut by enhancing mucosal integrity. PLoS ONE 2007;12:e1308.

45 Ghadimi D, Vrese MD, Heller KJ, Schrezenmeir J: Effect of natural commensal-origin DNA on Toll-like receptor 9 (TLR9) signaling cascade, chemokine IL-8 expression, and barrier integrity of polarized intestinal epithelial cells. Inflamm Bowel Dis 2010; 16:410–427.

Manfred Lamprecht, PhD, PhD
Centre for Physiological Medicine, Medical University of Graz
Harrachgasse 21/II, AT–8010 Graz (Austria)
Tel. +43 6641555528
E-Mail manfred.lamprecht@medunigraz.at, and
Green Beat – Institute of Nutrient Research and Sport Nutrition
Petersbergenstrasse 95b, AT-8042 Graz (Austria)
E-Mail office@greenbeat.at

Lamprecht M (ed): Acute Topics in Sport Nutrition.
Med Sport Sci. Basel, Karger, 2013, vol 59, pp 57–61

Pleuran (β-Glucan from *Pleurotus ostreatus*): An Effective Nutritional Supplement against Upper Respiratory Tract Infections?

Juraj Majtan

Institute of Zoology, Slovak Academy of Sciences, and Department of Microbiology, Faculty of Medicine, Slovak Medical University, Bratislava, Slovakia

Abstract

Prolonged and high-intensity exercise affects immune function and leads to an increased risk of upper respiratory tract infections (URTIs) in endurance athletes. The increased incidence of URTI symptoms may negatively affect athletic performance. Various nutritional supplements have been tested in the last decade for their ability to prevent developing of URTIs or reduce their incidence. One of the most promising nutritional supplements is β-glucan, a well-known immunomodulator with positive effects on functioning of immunocompetent cells. However, β-glucans are a diverse group of molecules that vary in macromolecular structure, solubility, viscosity, molecular weight and biological activity. This fact is supported by results from recent human clinical studies where β-glucans of different origin and properties differed in ability to prevent or reduce incidence of URTIs in athletes. It has been found that pleuran, a unique insoluble β-glucan isolated from mushroom *Pleurotus ostreatus*, significantly reduced the incidence of URTI symptoms in athletes. In addition, it was able to increase the number of circulating natural killer cells and to prevent reduction of natural killer cell activity. Contrarily, soluble oat β-glucan supplementation did not alter URTI incidence in endurance athletes. This difference suggests that the immunomodulatory capacity of β-glucans is strongly dependent on solubility and structural factors such as backbone structure and degree of branching. This review refers to using pleuran as a natural supplement that is able to protect endurance athletes against development of URTI.

It has been well known that prolonged and exhausting physical activity causes numerous changes in immunity and sometimes transient increases the risk of upper respiratory tract infections (URTIs). Changes in the immune system after heavy exercise are considered to be the main factors responsible for these infections [1]. The relationship between exercise and susceptibility to infection has been modeled in the form of a J-shaped curve [2]. This model assumes that the risk of URTIs increases with the intensity of the physical load. Several very recent studies have provided compelling evidence that high levels of physical activity are associated with increased risk of URTIs [3, 4]. On the other hand, the J-shaped model describing the relationship between exercise and susceptibility to infections is based on examination of symptoms associated to URTIs and not clinically diagnosed infections with a particular pathogen. Moreover, the direct dose-response relationship between exercise load and risk of URTI is not consistent and some studies suggest either no change, or even a slight reduction, in the risk of sickness after a run [5].

In general, prevention of URTI is always superior to treatment. One of the factors that could be responsible for elimination the risk of infection development is good nutrition. Endurance athletes must train hard for competition and are interested in strategies to keep their immune systems robust in order to avoid illness such as URTIs. They require additional nutritional supplements that are able to attenuate immune changes and inflammation following intensive exercise [6]. Some of these immunonutritional agents have already been proven in the general population where a reduction in the number, duration and severity of URTIs was observed. Thus, the influence of nutritional supplements on the immune response and incidence of URTI to heavy physical load is an interesting area of research [7–12].

β-Glucan as a Promising Nutritional Countermeasure

β-Glucans, glucose polymers, derived from a variety of sources including yeast, grain, or fungus, belong to a class of drugs known as biological response modifiers [13], and numerous studies have shown that (1,3)-β-D-glucans enhance immune functions such as anti-infective, antitumor and immunostimulatory activity [14–18]. Impressive effects in vitro have promoted further research in animal-based models. Rodent studies have indicated that oat β-glucan supplements offset the increased risk of infection associated with exercise stress [19–23]. Due to these proven beneficial properties, β-glucans have become attractive for athletes in endurance sports.

The effects of β-glucan supplementation on exercise-induced changes in immune function and upper respiratory tract symptoms have been investigated in four recent human studies of top-level athletes [24–27]. In a recent study, Nieman et al. [24] investigated the effect of an oat β-glucan supplement on exercise-induced suppression of the immune system in athletes. They found that oat-soluble β-glucan did not reduce the URTI incidence in cyclists during a 2-week period after intensive exercise. In contrast to this study, two other human studies showed that particulate β-glucan is able to reduce the incidence of URTI symptoms [25, 27]. These controversial findings suggest that the solubility and/or molecular structure of the β-glucans tested might determine their beneficial abilities such as reduction of URTI incidence and alternation of exercise-induced suppression of natural killer cell activity. It has been suggested that insoluble β-glucans, rather than their soluble forms, may play a role in preventing URTIs in athletes [27]. Furthermore, we also speculated that differences in biological

can further exacerbate these effects. *Exercise immunology* research has clearly and consistently demonstrated that many components of the immune system are temporarily reduced (immunodepression) after strenuous and/or prolonged bouts of exercise [3, 4]. This exercise-induced immunodepression may persist for as little as a few hours or a long as a few days. This depends largely on the nature of the exercise (in terms of intensity and duration). However, another key component is the frequency of exercise and/or recovery between bouts. For example, if subsequent bouts are commenced too soon, before the immune system has fully recovered, then a progressive accumulation of immunodepression may ensue resulting in chronically low immunity and increased infection risk. Periods of depressed immunity, whether small periods after individual sessions or more chronic periods, have been termed as 'open windows' during which athletes are more susceptible to picking up an infection. If an athlete should pick up an infection then this will undoubtedly have effects on performance – either by reducing performance, preventing an athlete from competing altogether and/or interfering with training. Obviously, this is detrimental and may cause considerable setbacks for athletes wishing to maintain a high level of performance and/or 'peak' for a specific competition. Although an appropriately structured and periodised training programme can go some way to minimising the risks, endurance athletes must train intensively if they are to be successful, and are consequently at an increased risk (compared to non-athletes) for considerable periods of time. In addition to the direct effects of exercise, other factors like psychological stress of competition, personal or life stresses, inadequate diet, lack of sleep, travel for competition etc., may have further negative effects on the immune system and thus resistance to infection. Therefore, taking measures to maximise the immune system (and minimise risk and exposure to infection) will be particularly beneficial to such athletes. It is worthy of further note at this point that whilst depressed immune function may increase the chances of picking up an infection, whether or not an athlete picks up an infection is also dependent on their exposure to pathogens, which may be beyond their control. Factors like having contact with lots of other people (e.g. crowded places, work, travel), increased ventilation during exercise, skin abrasions, environmental factors, etc., may also increase the exposure to or entry of pathogens into the body [3].

Bovine Colostrum

Bovine colostrum ('early milk') is the initial milk produced by cows, usually obtained within the first 48 h post-partum. Like 'normal' milk, it contains a rich source of nutrition, both in terms of macro- and micronutrients, but is also abundant in bioactive components including immune, growth and antimicrobial factors [5]. Bovine colostrum has received considerable attention within the *sports nutrition* and *exercise immunology* fields in recent years and has been suggested to confer immune, health and recovery benefits. A considerable amount of recent research has been conducted to investigate these claims. It is important to note here that all of the studies on (and reference to) bovine colostrum in this review refer to 'normal' bovine colostrum, from 'normal' dairy cows and *not* hyperimmune colostrum. For information only, hyperimmune bovine colostrum is obtained from cows that have been inoculated with or vaccinated against specific pathogens, meaning that the colostrum contains high concentrations of antibodies to a specific pathogen (or pathogens), usually for medical use. Hyperimmune colostrum is *not* considered here in relation to healthy athletes and active individuals. A number of studies have sought to determine whether daily supplementation with bovine colostrum, normally for periods of between 1 and 12 weeks, has a beneficial effect on immune function in athletes or individuals subjected to strenuous physical activity or intensive training regimes.

Bovine Colostrum Supplementation and Intestinal Barrier Integrity
Many stressors, such as heat stress, certain types of medication, and oxidative stress are known to cause disturbance to gut function and integrity. A number of these stressors may be induced by strenuous exercise (e.g. heat stress, oxidative stress), which has been shown to increase intestinal permeability. This increases permeability to luminal toxins and bacteria, allowing them into the systemic circulation (endotoxaemia). This may place additional stress on the immune system, which has to then deal with these bacteria and/or toxins, and this may be one factor contributing to exercise-induced immunodepression. However, increased intestinal permeability may also contribute to increased incidences of some acute gastrointestinal complaints and symptoms in athletes, such as stomach cramps, nausea and dizziness, and diarrhoea [6]. A high prevalence of such symptoms have been noted in some groups of athletes after strenuous competitive events (e.g. up to 50% incidence after marathon and ultra-endurance events or even >90% incidence during and after a very strenuous ultra-endurance event, an ironman triathlon [6]). Such symptoms have potentially negative effects on performance if incurred during an event, or subsequent recovery, dietary intake, training or performance if they develop afterwards. Under extreme circumstances (e.g. exhaustive exercise in a hot and humid environment) this can cause severe endotoxaemia, leading to acute inflammation, sepsis, shock and organ failure, which can be fatal in extreme situations (although this is rare). Indeed, it has been suggested that damage to the gut can be a contributing factor in exertional heat stroke [7]. Supplementation with bovine colostrum has been shown to be beneficial in maintaining intestinal integrity. For example, Marchbank et al. [8] conducted a placebo-controlled, counterbalanced crossover study to determine the effects of bovine colostrum supplementation on exercise-induced changes in intestinal permeability. Subjects were required to run on a treadmill for 20 min (which caused core body temperature to increase by between 1 and 2°C) after 2 weeks of daily supplementation with 20 g/day of bovine colostrum or a protein and energy-matched placebo. Running caused a significant increase in intestinal permeability in the placebo trial but this was almost completely blunted in the colostrum trial. Similar

beneficial effects of colostrum have been observed in human, animal and in vitro studies with cultured epithelial and gut cell lines (subjected to physical, chemical or thermal stressors). In summary, bovine colostrum supplementation may be beneficial in preventing exercise-induced increases in intestinal permeability and there is some preliminary evidence that this may be beneficial to athletes (e.g. by indirectly impacting upon training and performance), especially in those required to exercise or compete in hot (or unfamiliar) environments.

Bovine Colostrum and Immune Function

Bovine colostrum is of obvious importance for protection against infection in newborn calves and there is growing interest in this dairy product as a potential dietary supplement to counteract immunodepression in athletes. It should be noted, however, the mechanisms will be quite different in adult humans consuming bovine colostrum (as discussed in more detail below). That is, it is unlikely that there will be any passive transfer or that antibodies etc. from colostrum will survive digestion in the human gut. A more likely explanation is that small bioactive constituents (or their metabolites) appear after consumption, digestion and absorption and that these have direct effects on immune function (see below). Indeed, it has been demonstrated that bovine colostrum can enhance a number of human immune functions, from the functioning of peripheral blood leukocytes assessed in vitro [9–12], to in vivo immune functions such as the humoral immune response to an orally administered vaccine [13]. As discussed above, athletes may be at increased risk of infection during periods of strenuous training and stress and this may be detrimental to training and performance. Hence, there have been a number of recent studies that have investigated whether bovine colostrum is able to enhance immunity and/or reduce infections in athletes.

Mero et al. [14] observed a 33% increase in resting saliva secretory IgA (s-IgA) concentration after only 2 weeks of supplementation with 20 g/day bovine colostrum. Crooks et al. [15] recently demonstrated that a 12-week period of daily supplementation with a commercial chocolate drink powder (26 g/day) containing bovine colostrum (equivalent to ~12 g/day), in a group of distance runners, resulted in a 79% increase in resting saliva s-IgA concentration. The concentration and/or secretion of salivary s-IgA has been shown to be a good predictor of URT illness (or infection) risk in athletes, so it has been proposed that the increases observed with bovine colostrum may be beneficial and confer protection in athletes. Indeed, Brinkworth and Buckley [16] demonstrated a lower incidence of self-reported (with illness logs) URT illness symptoms in a group of men taking daily (60 g/day for 8 weeks) bovine colostrum compared to a placebo group. Other studies have shown that bovine colostrum supplementation improves other markers of immune function and/or prevents the magnitude of decrease following strenuous exercise (i.e. exercise-induced immunodepression), which may contribute to the protective effects in athletes. For example, Davison and Diment [17] observed that 4 weeks of daily bovine colostrum supplementation (20 g/day) prevented the prolonged exercise-induced decrease of salivary lysozyme concentration and secretion. They also assessed neutrophil degranulation capacity (stimulated degranulation) and although similar decreases in this marker of neutrophil function were seen immediately post-exercise (which was 2 h at ~64% of maximal oxygen uptake), by 1 h of recovery values were significantly higher in the colostrum group which could represent a more rapid closing of the 'open window'. These mechanisms may also contribute to the protective effects against illness in athletes. However, not all studies have observed beneficial effects on all measures and markers of immunity. For example, a recent study by Carol et al. [18] reported that bovine colostrum supplementation (25 g/day for 10 days) had no effect on 'immune variables' during a period of intensified training. However, the 'immune variables' measured in this study were limited to circulating concentrations of selected cytokines, hormones, blood cell counts and immunoglobulins (furthermore, none of the immunoglobulin concentrations changed as a result of the exercise). These measures are recognised as having relatively limited value as markers of human immune modulation [19] and functional measures (e.g. immune cell function rather than just counts) may be more valuable. Indeed, the work of Davison and Diment [17] found no effect on markers such as total or subset counts of blood leukocytes and cortisol concentration (in agreement with [18]) but that the functional measures (stimulated neutrophil degranulation) were enhanced post-exercise in the colostrum compared to the placebo group. Furthermore, Shing et al. [20] investigated the effects of bovine colostrum supplementation on salivary s-IgA in addition to some components of innate immunity. They observed some beneficial effects on some immune measures but not salivary s-IgA. Likewise, in the study of Davison and Diment [17], salivary s-IgA was not affected, in contrast to some of the previous studies showing enhanced s-IgA. Such findings highlight the fact that variations in research design, dosage (and perhaps quality of supplements) and duration of supplementation, type of exercise (e.g. does it even cause notable immunodepression in the first place), as well as variations between different individuals (e.g. type of athlete, possible non-responders) may have a large effect on the nature of any benefits provided by bovine colostrum for athletes [5] and these are factors requiring further attention in future research.

Mechanisms

In conjunction with their in vivo study, Marchbank et al. [8] used in vitro investigations with gut epithelial cells to demonstrate that bovine colostrum's effects on maintaining intestinal barrier integrity were attributable to a number of mechanisms that can ultimately result in better maintenance of tight junctions under thermal, and possibly oxidative, stresses. The observed reduction of heat-induced apoptosis in these cells was shown to be partly attributed to epidermal growth factor (EGF) present in the colostrum, confirmed by incubation with anti-EGF antibodies, although some of the anti-apoptotic processes were unaltered by anti-EGF, showing that other factors must also contribute. Marchbank et al. [8] also showed that bovine colostrum enhanced heat-shock proteins induction under basal and heat stress conditions.

This is another possible mechanism to explain the beneficial effects of colostrum as heat-shock proteins provide cellular protection against numerous insults, including heat and oxidative stresses, and it is likely that this also contributed to the truncated exercise-induced rise in intestinal permeability seen in this study.

The underlying mechanism responsible for the effect of bovine colostrum on human immune parameters remains unclear. One possible, indirect, mechanism is by reducing the exercise-induced increase in intestinal permeability as discussed above. For example, increased intestinal permeability allows the translocation of luminal bacteria (or toxins) into the systemic circulation, which may place additional stress on the immune system and/or may present an additional point of entry for infection causing pathogens. Therefore, the effects of bovine colostrum on maintaining intestinal integrity may contribute to the effects on immune function after strenuous exercise. In addition, evidence from in vitro cell culture studies suggests that bioactive, low molecular weight (≤10 kDa) components (such as the proteose peptones) may enhance leukocyte capacity by a direct effect [11, 12]. However, if this is the case, then it would be expected that the effects would be apparent after acute supplementation (immediately after the absorption of these components). To test this hypothesis, we carried out a number of studies [21, 22] to assess the effects of acute ingestion. In the first study [21] we observed an increase of between 25 and 35% in neutrophil-stimulated oxidative burst capacity with colostrum compared to placebo ingestion (30 g) at rest, and this persisted for at least 3 h. We also demonstrated that acute supplementation had some beneficial effects on the exercise-induced decrease in neutrophil-stimulated oxidative burst subsequent to prolonged (2.5 h) cycling exercise [22]. Taken together with the in vitro work previously mentioned [10–12], this does provide evidence in support of the idea that the mechanisms are associated with bioactive components that are biologically available to effect immune functions after ingestion of bovine colostrum by humans. Furthermore, when plasma was isolated from a subject who had consumed bovine colostrum or placebo (1 and 3 h post-consumption), the addition of that plasma to pre- and post-exercise blood samples in the same subject (but on a subsequent occasion) also enhanced neutrophil-stimulated oxidative burst and reduced the exercise-induced decrease [21]. This provides further evidence that biologically available constituents, after ingestion, are at least partly responsible for the effects on these immune functional measures in exercising subjects. However, somewhat surprisingly, the effects observed in the acute ingestion studies do not appear to be as great as those observed with longer-term supplementation [17], meaning that there is likely to be some other mechanisms contributing to the greater effects seen in such studies (i.e. >2 weeks supplementation periods). However, these mechanisms remain elusive at present and further research is required before we can fully determine all mechanisms involved. Despite this, the existing evidence does support the notion that bovine colostrum is beneficial for athletes involved in strenuous training (e.g. endurance athletes) in terms of immunity and resistance to infection, which may ultimately be beneficial to performance.

Conclusion

To conclude, it is well known that strenuous and/or prolonged exercise causes transient perturbations in immune function. However, athletes must train intensively if they are to be successful, and consequently are at an increased risk (compared to non-athletes) of picking up infections for considerable periods of time and taking measures to maximise the immune system will be particularly beneficial to such athletes. Bovine colostrum supplementation has been investigated as a possible nutritional countermeasure to enhance (or maintain) immune function following strenuous or prolonged exercise or during intensive training periods. There is convincing evidence that daily supplementation with bovine colostrum, for a number of weeks (and preliminary evidence for acute effects after a single dose), can maintain intestinal barrier integrity, immune function and reduce the chances suffering URT infections or symptoms in athletes or those undertaking heavy training. Preliminary work suggests that the beneficial effects on immune function are attributable, at least in part, to small bioactive components that survive digestion and are biologically available after consumption. However, further work is required to determine all of the mechanisms.

Disclosure Statement

The author has no conflicts of interest to disclose.

References

1 Nieman DC: Exercise, upper respiratory tract infection, and the immune system. Med Sci Sports Exerc 1994;26:128–139.
2 Pedersen BK, Kappel M, Klokker M, Nielsen HB, Secher NH: The immune system during exposure to extreme physiologic conditions. Int J Sports Med 1994;15(suppl):116–121.
3 Gleeson M: Immune function in sport and exercise. J Appl Physiol 2007;103:693–699.
4 Davison G, Simpson RJ: Immunity; in Lanham-New SA, Stear SJ, Shirreffs SM, Collins AL (eds): Sport and Exercise Nutrition. Chichester, Wiley-Blackwell, 2011, pp 281–303.
5 Shing CM, Hunter DC, Stevenson LM: Bovine colostrum supplementation and exercise performance: potential mechanisms. Sports Med 2009;39: 1033–1054.
6 Jeukendrup AE, Vet-Joop K, Sturk A, Stegen JH, Senden J, Saris WH, Wagenmakers AJ: Relationship between gastrointestinal complaints and endotoxaemia, cytokine release and the acute-phase reaction during and after a long-distance triathlon in highly trained men. Clin Sci (Lond) 2000;98:47–55.
7 Leon LR, Helwig BG: Heat stroke: role of the systemic inflammatory response. J Appl Physiol 2010; 109:1980–1988.
8 Marchbank T, Davison G, Oakes JR, Ghatei MA, Patterson M, Moyer MP, Playford RJ: The nutriceutical, bovine colostrum, truncates the increase in gut permeability caused by heavy exercise in athletes. Am J Physiol Gastrointest Liver Physiol 2011; 300:G477–G484.
9 Biswas P, Vecchi A, Mantegani P, Mantelli B, Fortis C, Lazzarin A: Immunomodulatory effects of bovine colostrum in human peripheral blood mononuclear cells. New Microbiol 2007;30:447–454.

10 Shing CM, Peake JM, Suzuki K, Jenkins DG, Coombes JS: Bovine colostrum modulates cytokine production in human peripheral blood mononuclear cells stimulated with lipopolysaccharide and phytohemagglutinin. J Interferon Cytokine Res 2009;29:37–44.

11 Sugisawa H, Itou T, Ichimura Y, Sakai T: Bovine milk enhances the oxidative burst activity of polymorphonuclear leukocytes in low concentrations. J Vet Med Sci 2002;64:1113–1116.

12 Sugisawa H, Itou T, Saito M, Moritomo T, Miura Y, Sakai T: A low-molecular-weight fraction of bovine colostrum and milk enhances the oxidative burst activity of polymorphonuclear leukocytes. Vet Res Commun 2003;27:453–461.

13 He F, Tuomola E, Arvilommi H, Salminen S: Modulation of human humoral immune response through orally administered bovine colostrum. FEMS Immunol Med Microbiol 2001;31:93–96.

14 Mero A, Kähkönen J, Nykänen T, Parviainen T, Jokinen I, Takala T, Nikula T, Rasi S, Leppäluoto J: IGF-I, IgA, and IgG responses to bovine colostrum supplementation during training. J Appl Physiol 2002;93:732–739.

15 Crooks CV, Wall CR, Cross ML, Rutherfurd-Markwick KJ: The effect of bovine colostrum supplementation on salivary IgA in distance runners. Int J Sport Nutr Exerc Metab 2006;16:47–64.

16 Brinkworth GD, Buckley JD: Concentrated bovine colostrum protein supplementation reduces the incidence of self-reported symptoms of upper respiratory tract infection in adult males. Eur J Nutr 2003;42:228–232.

17 Davison G, Diment BC: Bovine colostrum supplementation attenuates the decrease of salivary lysozyme and enhances the recovery of neutrophil function after prolonged exercise. Br J Nutr 2010; 103:1425–1432.

18 Carol A, Witkamp RF, Wichers HJ, Mensink M: Bovine colostrum supplementation's lack of effect on immune variables during short-term intense exercise in well-trained athletes. Int J Sport Nutr Exerc Metab 2011;21:135–145.

19 Albers R, Antoine JM, Bourdet-Sicard R, et al: Markers to measure immunomodulation in human nutrition intervention studies. Br J Nutr 2005;94: 452–481.

20 Shing CM, Peake J, Suzuki K, Okutsu M, Pereira R, Stevenson L, Jenkins DG, Coombes JS: Effects of bovine colostrum supplementation on immune variables in highly trained cyclists. J Appl Physiol 2007;102:1113–1122.

21 Davison G, Thatcher R, Jones A: Acute bovine colostrum supplementation enhances neutrophil oxidative burst at rest and following immunodepressive exercise: a pilot study. Proc 10th Symp International Society of Exercise and Immunology, Oxford 2011.

22 Jones A, Thatcher R, Davison G: The effect of acute bovine colostrum supplementation on neutrophil responses to prolonged cycling. Proc Annual Congress of the European College of Sport Science, Liverpool 2011.

Glen Davison
Endurance Research Group
University of Kent, School of Sport & Exercise Sciences
Chatham ME4 4AG (UK)
Tel. +44 01634 888994, E-Mail G.Davison@kent.ac.uk

Lamprecht M (ed): Acute Topics in Sport Nutrition.
Med Sport Sci. Basel, Karger, 2013, vol 59, pp 70–85

Supplementation with Mixed Fruit and Vegetable Concentrates in Relation to Athlete's Health and Performance: Scientific Insight and Practical Relevance

Manfred Lamprecht

Institute of Physiological Chemistry, Centre for Physiological Medicine, Medical University of Graz, and Green Beat, Institute for Nutrient Research and Sport Nutrition, Graz, Austria

Abstract

Regular consumption of fruits and vegetables (FV) is widely regarded as an important contributor to a healthy diet. Inadequate consumption of plant foods is associated with an inadequate supply of important micronutrients like vitamins, phytochemicals and minerals. In athletes a deficit of these micronutrients can lead to excessive production of reactive oxygen and nitrogen species that induce tissue damage, a higher frequency of inflammatory processes, decreased immunity, increased susceptibility to injury, and prolonged recovery. But many athletes rarely achieve the recommended intake of FV due to difficult coordination of training activities and food intake, or due to problems with digestion of FV. Therefore, in recent years more and more sports people have adopted supplemental FV concentrates to work around timing problems with uptake and the detrimental digestive effects during training of high FV intake. It is thought that supplementation of an athlete's basic diet with mixed FV concentrates can promote stable health and immunity, in order to provide a basis for optimal adaptation and performance. The intention of this article is to build a bridge between the science behind FV supplementation in exercise on the one hand and the practical relevance of its application on the other. For that purpose this paper addresses three questions: Is supplementation with a mixed FV concentrate to the athlete's diet appropriate to ensure stable health and immunity? Can supplementation with a mixed FV concentrate improve performance? Counseling guidance: how can sport nutrition advisors decide whether or not to supplement with mixed FV concentrates?

The health benefits of regular consumption of fruits and vegetables (FV) are well established. A variety of scientific publications demonstrate that adequate consumption of plant foods is associated with a decreased risk of chronic degenerative diseases, such as coronary heart disease, stroke, diabetes, or certain types of cancer [1–12]. The risk-reducing effects are attributed to bioactive components including phytochemicals, phytonutrients and vitamins, minerals, and fiber [13–16].

Internationally, various Public Health Nutrition strategies such as '5 a day' are used to try to encourage people to increase consumption of FV. However, these have met with limited success: nutrition reports and surveys from USA, England, Germany, and many other countries reveal that people consume about 300 g of FV per day [17–22], far less than the recommended 400 up to 650 g/day [23–25].

Many athletes rarely achieve recommended intake of FV due to difficult coordination of training units and food intake. FV consumption before exercise training can also lead to digestive discomfort during exercise due to high fructose and fiber content [26, 27]. Inadequate FV intake in an athlete's daily diet can lead to inadequate consumption of bioactive compounds. This situation provides a rationale for offering concentrated FV nutrition, with less fiber and fructose to this target group.

The well-balanced mixture of phytonutrient vitamins, minerals, and other bioactives from a variety of FV may lead to additive and synergistic interactions in human metabolism that result in health benefits [28, 29]. Hence, to bring as many as possible of these FV-bioactives together in one supplement might be superior to supplements containing only vitamins or juice or powder from just one or a few fruits and/or vegetables.

This article refers to studies that used supplementation with mixed FV concentrates in relation to exercise and performance sport. A search of Medline and the Cochrane Library returned publications on a single preparation (Juice Plus+®; NSA, Collierville, Tenn., USA). This product is a whole food-based encapsulated mixed fruit, vegetable and berry concentrate recently reviewed [30]. To date there are five research articles that pertain to exercise and performance sport and one pilot sports study on the same product conducted in our own laboratories.

Supplementation with Mixed Fruit and Vegetable Juice Concentrates to Athlete's Nutrition: Health Benefits or Not?

Increased metabolism due to exercise training results in enhanced demands for energy, protein, carbohydrate, water, essential fatty acids, and also micronutrients such as vitamins, phytonutrients and minerals. A deficit of micronutrients with antioxidant functions can result in an imbalance in the athlete's redox biology in favor of accumulation of reactive oxygen and nitrogen species (RONS) and disturbed redox signaling and control. This situation is called oxidative stress, resulting in molecular, cell and tissue damage [31–34]. In athletes, an overwhelming production of RONS can induce increased inflammatory processes, decreased immunity, increased susceptibility to injury and prolonged recovery [35–38].

To stabilize health and to avoid oxidative stress and its consequences, increased consumption of FV is prudent. However, there is no generally accepted recommendation of FV consumption for athletes, although some institutions and authors recommend up to 13 portions a day in proportion to energy expenditure [39].

The standard recommendation to consume 5 portions of FV per day is already difficult to achieve for the general population [40]. For athletes, in many cases, consuming a high volume of plant foods is not realistic. With reference to national nutrition campaigns in Austria [41] and Germany [42], 1 portion of uncooked vegetables and fruits weighs 100–200 g and 1 portion of cooked vegetables weighs 200– 300 g. In total, this leads to a minimum weight of 500 g/day and an average weight of 750 g of FV per day. Very often, top athletes conduct training regimens with two or more training units per day. On these days, consuming more than 3 portions or 400 g of fruits and/or vegetables, with meals/snacks before or within training sessions, might cause digestive issues, especially due to the high content of fiber in FV. Therefore, from the scientific as well as the practical point of view, it makes sense to search for alternatives that can – at least in part – compensate inadequate consumption of plant foods in an athlete's basic diet.

Mixed FV juice concentrates have been on the market since 1993. They focus primarily on a normal population who feel they do not eat enough plant foods on a regular basis. In recent years, athletes have also adopted these products to circumnavigate the detrimental digestive effects of high FV intake, while getting some of the beneficial effects. Supplementation of an athlete's basic diet with mixed FV concentrates may make a contribution to promote stable health and immunity, and therefore provide support for optimal adaptation and performance.

Scientific literature is scarce regarding athletic people with respect to supplementation with mixed FV concentrates, although a recent systematic review included research on healthy subjects – both trained and untrained [30]. This review revealed that daily consumption of the commercially available encapsulated mixed fruit and vegetable concentrates increased serum concentrations of major antioxidant vitamins. Esfahani et al. [30] also reported reduced concentrations of oxidative stress and inflammatory markers and promising health advantages on immunity and endothelial function. They noted a diversity of studies with respect to design, study population and the variability in the measured outcomes and assays utilized.

A pilot study with trained mountain bikers, conducted in our own laboratories, revealed that a 3-week supplementation with a mixed FV juice concentrate reduced concentrations of carbonyl proteins (CP) – a marker of RONS-induced protein oxidation – before and after strenuous mountain biking. We also observed reduced pre- and post-exercise concentrations of malondialdehyde – a marker of lipid oxidation [43]. The mixed FV juice concentrate was given with the last meal before exercise and with the first meal post-exercise.

Bloomer et al. [44] reported that use of the fruit, vegetable and berry (FVB) form of this supplement resulted in reduced exercise-induced increase of plasma protein carbonyl concentrations – a marker of RONS-induced protein oxidation – compared to placebo after 30 min treadmill running at 80% of $Vo_{2\ max}$. Trained subjects had consumed the nutraceutical continuously for 2 weeks with their meals.

This group also conducted a gender comparison of exercise-induced oxidative stress [45]. They found that trained women had higher resting antioxidant levels than their trained male counterparts. Markers of oxidative stress increased similarly in both genders in response to exercise of similar intensity and duration. They concluded that 2 weeks of supplementation with the FVB concentrate can attenuate exercise-induced oxidative stress equally in both genders.

In a randomized, double-blinded, placebo-controlled trial in a cohort of trained men [46], all non-smokers, we demonstrated that daily supplementation for 28 weeks with the FVB capsules (with meals) again reduced CP. In this study, we also found a positive effect of FVB supplementation on subject's immunity expressed via reduced frequency of common cold, sore throat, and fever. This effect was even more pronounced when subjects' duty became more stressful, due to more hours of work and circadian imbalance. In this stressful time period, we also found significantly lower concentrations of inflammatory cytokine TNF-α in the supplemented group compared to placebo. No adverse effects of supplementation were observed. However, it is noteworthy that the investigated group only consumed 2–3 portions of FV per day, far below recommendations. We believe that was the main reason why the supplementation with the mixed FVB concentrate could demonstrate beneficial results.

With the same cohort of trained men we also conducted endurance tests with distinct intensities: at 70% of $Vo_{2\,max}$ and at 80% of $Vo_{2\,max}$ [47]. The 70% intensity was adjusted about 10% below anaerobic threshold the second slightly above. At each intensity level, we tested both a placebo group and the active group, which received the encapsulated mixed FVB concentrate. We found that post-exercise CP concentrations significantly increased at 80% $Vo_{2\,max}$ intensity but not at 70%, but this phenomenon only occurred in the placebo group. Towards the end of the study (28 weeks), when individual stress profile was increased by 45% more hours on duty per week, CP concentrations approached 1 nmol/mg protein after 80% $Vo_{2\,max}$ intensity in the placebo group. Referring to our own laboratory data obtained from athletes during the last years, CP concentrations close to 1 nmol/mg protein (based on our applied method described in ref. 47) are related to increased events of common cold and inflammation (unpubl. data).

In summary, we observed the following differences between groups supplemented with either FVB or placebo: (1) increased resting CP and TNF-α values in the placebo groups towards the end of the study (16 and 28 weeks); in this time period individual stress profile was significantly increased; (2) in the last 20 weeks of the study a tendency to more duty days lost due to illness in the placebo group (p = 0.068); (3) after strenuous endurance exercise (80% of $Vo_{2\,max}$) a significant increase in CP only in the placebo group, and (4) this increase in CP approached 1 nmol/mg protein towards the end of the study, after 28 weeks in the placebo group (table 1).

Additionally it is noteworthy to mention that FVB supplementation showed neither prooxidant effects nor changes of antioxidant enzymes in any of our studies [43, 46, 47]. This is in contradiction to some exercise studies which have reported

Table 1. Differences between long-term FVB supplementation and placebo in post-exercise CP concentrations (compared to pre-exercise) of stressed, non-smoking, trained men [modified from Lamprecht et al., 2009]

Weeks with FVB/placebo	FVB 70% $VO_{2\,max}$	FVB 80% $VO_{2\,max}$	Placebo 70% $VO_{2\,max}$	Placebo 80% $VO_{2\,max}$
0 baseline	=	↑	=	↑
4 weeks	=	=	=	↑
16 weeks	=	=	=	↑
28 weeks	=	=	=	↑ (~1 nmol/mg)

increased lipid peroxidation and decreased plasma glutathione peroxidase (GPx) in trained men after antioxidant supplementation [48–50]. A decrease of antioxidant enzyme GPx could cause a weakening of the body's antioxidant system. This means that supplementation with antioxidants does not provide a net benefit when internal antioxidant systems are regulated down in parallel. Obviously this downregulation does not occur with adequate supplementation with mixed FVB concentrates, even after a long-term supplementation of 7 months.

Recently, Goldfarb et al. [51] showed that 4-week supplementation with the same mixed FVB juice concentrate (with meals) leads to significantly lesser increases in CP, MDA and oxidized glutathione after eccentric exercise. They found no differences between supplementation and placebo in the context of functional changes related to pain and muscle damage between their non-resistance trained study subjects.

Nevertheless, this review of existing scientific data gives some evidence that supplementation with mixed FVB concentrates to the athlete's diet can provide beneficial effects. To summarize, data from studies with FVB supplementation indicate that an athlete's immunity and health can benefit when (1) dietary consumption of FV is low (<3 portions a day), (2) an increased psychological stress profile occurs and – given one or both factors – (3) during that time strenuous exercise bouts/training units are conducted (table 2).

Can Supplementation with Mixed FVB Concentrates Improve Performance?

For sport nutrition experts and counselors in this field it is self-evident that one has to discuss the role of mixed FVB supplements in the context of ergogenic effects. PubMed reveals only one study that investigated a direct functional outcome of mixed FVB supplementation on performance in untrained young men [51]. Goldfarb et al. could not find an influence of a 28-day FV supplementation/pretreatment on the range of motion or maximal isometric muscle force after one acute bout of eccentric exercise. Nevertheless, they found increased concentrations of oxidative stress parameters post-exercise in the placebo group but not in the FVB-supplemented group. It could be interesting to observe comparisons between groups when other exercise protocols are applied with stronger emphasis on the recovery period. Repeated bouts of eccentric exercise within a defined timeline might repeatedly reveal RONS accumulation and oxidative stress due to muscle damage. Such exercise protocols with short recovery periods either with or without specific nutritional regimens and target groups, for example the elderly, would contribute to clarify this issue.

Unfortunately, no studies could be identified on PubMed regarding mixed FVB supplementation and its influence on performance in trained people. As the mixed FVB concentrate used in our studies (Juice Plus+®) is standardized with antioxidant vitamins E, C and β-carotene, it is valid to refer to studies which investigated mixtures of antioxidants as ergogenic aids and against muscle damage.

There is some evidence that antioxidant supplementation could offer protection from exercise-induced muscular and oxidative damage, inflammation, muscle force loss and fatigue [52–57]. If so, this would accelerate recovery, especially from resistance training, and consequently lead to increased strength performance. On the other hand, a number of studies suggest that antioxidant supplementation might promote muscle damage and hinder recovery [58–61]. These conflicting data are due to the diversity of study protocols with different methods, subjects, surrogate endpoints, outcome measures, products, etc.

Many studies that investigated the direct ergogenic role of antioxidant supplements in endurance- and resistance-trained men have failed, especially when mixtures of the classical antioxidants E, C and coenzyme Q10 were applied [62–67]. Researchers postulate that antioxidant supplements could hinder the beneficial cell adaptations to exercise via RONS-induced signal transduction. Some studies showed [68–70] that antioxidant supplementation induced decreased activation of protein kinases, followed by blunted DNA binding of transcription factors. These mechanisms result in reduced gene expression of antioxidant enzymes. Gomez-Cabrera et al. [71] observed that antioxidant supplementation inhibited upregulation of antioxidant enzymes GPx and superoxide dismutase in animal muscles.

Interestingly, some recent studies with polyphenol-containing supplements including quercetin, *Rhodiola rosea* or beetroot juice revealed performance enhancing effects in trained cyclists or rowers [72–74]. FVB supplements include a lot of polyphenol-containing plants, which is an interesting aspect for future exercise research.

There is emerging evidence that the antioxidant potential of phenolic compounds is unlikely to be the sole mechanism responsible for the beneficial effects. Interaction with various key proteins in cell signal transduction cascades is described [75].

Although no investigation with mixed FVB supplements on exercise performance of trained people is published at this time, one can assume a low probability that these supplements will provide direct ergogenic effects. It appears to be more useful to investigate these supplements under the perspective of nutraceuticals in regard to their indirect effects on performance: e.g. recovery from strenuous and repetitive

Table 2. Summary of sport studies with mixed FV or FVB supplements

Dosage of supplementation	Subjects' characteristics	Time points of sample collection
Fruit and vegetable juice concentrate: 2 × 2 capsules a day for 3 weeks	5 trained male mountain bikers; 32–42 years; non-smokers	Blood collection before exercise, immediately post-exercise and 1 h post-exercise
Fruit, vegetable and berry juice powder conc.: 2 × 3 capsules a day for 2 weeks; 400 IU vitamin E + 1,000 mg vitamin C for 2 weeks	Trained men and women: 25 men, 23 women; 18–30 years; non-smokers	Blood collection at baseline, after 2 weeks of FVB suppl., after 1 week washout; also before exercise and immediately post-exercise
Fruit, vegetable and berry juice powder conc.: 2 × 3 capsules a day for 28 weeks	41 trained men; 30–40 years; non-smokers	Blood collection at baseline, 4, 16, and 28 weeks of suppl. or placebo; resting values
Fruit, vegetable and berry juice powder conc.: 2 × 3 capsules a day for 2 weeks; 400 IU vitamin E + 1,000 mg vitamin C for 2 weeks	Trained men and women: 25 men, 23 women; 18–30 years; non-smokers	Blood collection at baseline, after 2 weeks of FVB suppl., after 1 week washout; also before exercise and immediately post-exercise
Fruit, vegetable and berry juice powder conc.: 2 × 3 capsules a day for 28 weeks	41 trained men; 30–40 years; non-smokers	Blood collection at baseline, 4, 16 and 28 weeks of suppl. or placebo; also before exercise and immediately, 30 min and 30 h post-exercise
Fruit, vegetable and berry juice powder conc.: 2 × 3 capsules a day for 28 days + 4 days post-exercise	44 non-resistance trained subjects; no gender specification; 18–35 years; non-smokers	Blood collections and muscle function tests before exercise and immediately, 2, 6, 24, 48 and 72 h post-exercise

Design methodology	Results outcome	Author year
Pilot study; field study: 75% $VO_{2\,max}$ mountain bike exercise for 2 h followed by 45% $VO_{2\,max}$ exercise for 1 h; trial was conducted at 2 times: 1st time without, 2nd time after 3 weeks of FV supplementation	After 3 weeks of FV supplementation decreased MDA and CP concentrations before, after 2 h at 75% $VO_{2\,max}$ and after an additional h at 45% $VO_{2\,max}$ mountain bike exercise; no influence of the exercise model on CP and MDA	Lamprecht et al., 2005
Double-blind, randomized, placebo-controlled, 3 parallel groups: 2 supplemented groups and 1 placebo group; controlled diet before exercise tests; 30 min running exercise at 80% $VO_{2\,max}$ before and after 2 weeks supplementation/placebo and after 1 week washout	Reduced exercise-induced increase of plasma PC in FVB-supplemented groups; no differences in PC between supplemented groups; no impact of supplementation on MDA and 8-OHdG	Bloomer et al., 2006
Double-blind, randomized, placebo-controlled, 2 parallel groups: supplemented and placebo group; standardized nutrition	Plasma CP lower in the supplemented group; differences between groups more distinct with increased stress profile; tendency to fewer duty days lost due to illness (p = 0.068) in the supplemented group	Lamprecht et al., 2007
Double-blind, randomized, placebo-controlled, 3 parallel groups: 2 supplemented groups and 1 placebo group; controlled diet before exercise tests; 30 min running exercise at 80% $VO_{2\,max}$ before and after 2 weeks suppl./placebo and after 1 week washout; gender comparison of exercise-induced oxidative stress	Women have higher resting antioxidant levels than men; markers of oxidative stress increased similarly in both genders in response to exercise of similar intensity and duration; 2 weeks of antioxidant suppl. can attenuate exercise-induced oxidative stress equally in both genders	Goldfarb et al., 2007
Double-blind, randomized, placebo-controlled, 2 parallel groups: supplemented and placebo group; standardized nutrition; 40 min cycle ergometer exercise at 70% and 80% of $VO_{2\,max}$ at baseline, 4, 16 and 28 weeks	Increase in CP in placebo group after 80% $VO_{2\,max}$ exercise; post-exercise CP differences more distinct with increased stress profile; no influence of supplementation on antioxidant enzyme activities and redox state of albumin	Lamprecht et al., 2009
Double-blind, randomized, placebo-controlled, 2 parallel groups: supplemented and placebo group; documented diet before exercise tests; 4 sets of eccentric elbow flexion with non-dominant arm	Muscle soreness, range of motion and maximal isometric force not influenced by treatment; PC, MDA and glutathione ratio increased post-exercise in placebo group whereas no changes in the supplemented group; no influence of treatment on lipid peroxides	Goldfarb et al., 2011

resistance training, regeneration from injury and traumata, or general stabilization of athlete's immunity.

Inevitable: Athlete's Health and Performance Are Housed under the Same Roof!

It is notable that stabilized immunity has the largest impact on athlete's performance. In competition periods with high intensities and (too) short recovery periods, stressful general conditions, and a higher risk to suffer oxidative stress, inflammation and imbalanced immunity occurs. During such periods stabilization of immunity has priority and a possible non-favorable effect on adaptation is secondary. There is already evidence that FVB supplementation has beneficial effects on oxidative stress, inflammation and immunity in trained and untrained people [44, 46, 47, 76]. These effects seem to increase when basic nutrition is lacking in plant food consumption [46].

Gomez-Cabrera [71] demonstrated the beneficial effect of restricted RONS production in competitive periods via cyclists taking part in the Tour de France: when given allopurinol (a xanthine oxidase inhibitor and antioxidant), they had lower increases in the activity of creatine kinase and aspartate aminotransferase.

Once an athlete becomes sick, recovery from common cold in top endurance sports takes at least 6–8 weeks to reach top form again. In cross-country skiing for example, this would mean a lost season if such a disease event occurred around Christmas time. As long as the athlete can keep his health, he/she stays competitive. Moreover, in competition periods athletes usually are not focused on performance improvement. The time period in which the highest level of fitness is reached has to be programmed prospectively and is already achieved when the competition season begins.

Consequently, it is useful to base one's decision regarding FV supplementation on the primary goal: adaptation to a programmed exercise stress or stabilization of immunity and health? In preparation and development periods, supplementation with antioxidant-containing FV supplements might be counterproductive due to hindered RONS-induced signal transduction. In competitive periods, the supplementation is beneficial to stabilize immunity, if adequate testing on the product demonstrated these immune stabilizing effects. But also other domains contribute to a pro or con decision as described in the next section.

How Can Sport Nutrition Advisors Orientate to Decide Pro or Contra Supplementation with FVB Concentrates?

To decide pro or con regarding FVB supplementation to an athlete's diet, at least five domains have to be considered:

(1) The athlete's typical diet (D): A 7-day food record is necessary to estimate the basic diet and also daily plant food intake. A non-favorable characteristic of this

Table 3. An overview and examples for non-favorable or favorable manifestations of each influencing domain

Favorable	Non-favorable
Diet (D): – Consumption of ≥5 portions of FV per day – Balanced mixed basic diet – Periods with normal energy uptake (most likely determined by influencing domain T) – Consumption of >5 portions of FV in periods of high energy expenditure – Consumption of other food supplements or specific functional food	Diet (D): – Consumption <5 portions of FV per day – Diets with undersupply of micronutrients especially antioxidants (e.g. severe weight loss practice, unbalanced diet, etc.) – Periods of high energy uptake (most likely determined by domain T) and consumption of <6–13 portions of FV per day – Consumption of other food supplements or specific functional food
Biochemistry and anamnesis (B): – Oxidative stress, inflammatory parameters or antioxidants in blood and/or urine within appropriate ranges before and post-exercise – Appropriate concentrations of parameters of the standard blood chemistry panel – Beneficial genetic constitution – Inconspicuous anamnesis	Biochemistry and anamnesis (B): – Oxidative stress + inflammatory parameters + antioxidants in blood and/or urine out of appropriate ranges before and post-exercise – Inappropriate standard blood chemistry panel (e.g. imbalance in lipid metabolism) – Unfavorable genetic constitution – Unfavorable anamnesis
Training (T): – Low intensity and duration – Low frequency, adequate recovery periods – Periods with strenuous training units to achieve optimized performance with adequate time for recovery/super compensation	Training (T): – Strenuous competition period – Rehabilitative or weight management training – Periods with high intensity and frequency combined with inevitable short recovery periods and high energy expenditure
Basic conditions (C): – Favorable and relaxed psycho-physiological situation – Good environment, facilities, equipment for training, favorable humidity...	Basic conditions (C): – Significant increased psycho-physiological stress profile – Inappropriate facilities, environment and humidity, stay in high elevations, pollutants etc.
FV product (P): – Documented scientific evidence for bioefficacy of the product (data/publications from tests on the product) – Good quality: safety and risk assessment certificates available	FV product (P): – No scientific evidence for bioefficacy of the product (no testing on the product) – High concentrations of antioxidant vitamins added to the product – No safety and risk assessment certificates

domain indicates supplementation with FVB concentrates, e.g. individual digestive situation or certain aversions/preferences do not allow dietary intake of 5 portions a day or more. For examples of favorable or non-favorable characteristics of the five influencing domains, see table 3.

(2) Biochemical checkup and clinical history (B): Analysis of oxidative stress and inflammatory parameters in blood and urine as well as a standard blood chemistry

panel. With respect to an athlete's history, age, gender, and genetics, it is useful to conduct a panel of several parameters at rest and post-exercise. FVB supplementation is a tool to combat against excess oxidative stress or inflammation.

(3) Characteristic of exercise training and competition (T): In certain time periods, stabilization of immunity and avoidance of inflammatory processes are of the highest priority. During this time, FVB supplementation is indicated. During macrocycles primarily focused on adaptation, it might be indicated to reduce or even avoid supplementation with mixed FVB concentrates.

(4) Underlying 'basic conditions' in an athlete's training period (C): Training in hot and humid weather, stay in higher regions, cold and dry climate, polluted air, unfavorable training facilities, etc. Also stressful daily life, mentally/emotionally disturbed everyday life, job, family, etc. fall into this domain and contribute to decide pro or con FVB supplementation.

(5) Quality and evidence of the FVB supplement/product itself (P): If FVB supplementation is indicated the product has to provide scientific evidence of bioefficacy and must fulfill all safety criterions. Well-described methodology, evaluated dosage, timing and investigated target groups are key factors necessary to decide to choose a product. Possible adverse effects and recommendations to avoid digestive problems should be published to give opportunity for quick adjustment.

Each domain should be evaluated for advantageousness on the athlete's health and performance as well as all five domains among each other. In the event that one or more certain domains reveal non-favorable characteristics/manifestations, the likelihood for FVB supplementation increases (except for P, cf. fig. 1). Figure 1 provides a 'domain grid' to ease the decision pro or con FV supplementation to an athlete's diet.

If FVB supplementation is indicated the product should be added to one or more meals, depending using the regimen from published scientific studies with that product. The decision pro or con regarding FVB supplementation has to be reevaluated continuously. Based on the subject's genetic profile, absorption rates, bioavailability, pharmacokinetics, bioactivity, etc. might be different among people and could lead to different responses, even to supplementation regimens with scientific evidence for efficacy. Thus, after 3–16 weeks, depending on training macrocycle and/or period of the specific type of sport, frequently conducted counseling sessions with checkups comprising all influencing domains reflect the most professional handling with this issue.

Note: Supplements are no replacement for dietary fiber! Even if FVB supplementation is indicated, these supplements cannot substitute for of the recommended daily fiber intake of 20–35 g [77]. Especially in less strenuous training periods (T), the uptake of dietary plant foods should increase to at least 5 portions a day as long as this amount and frequency of intake does not result in digestive problems.

Lamprecht

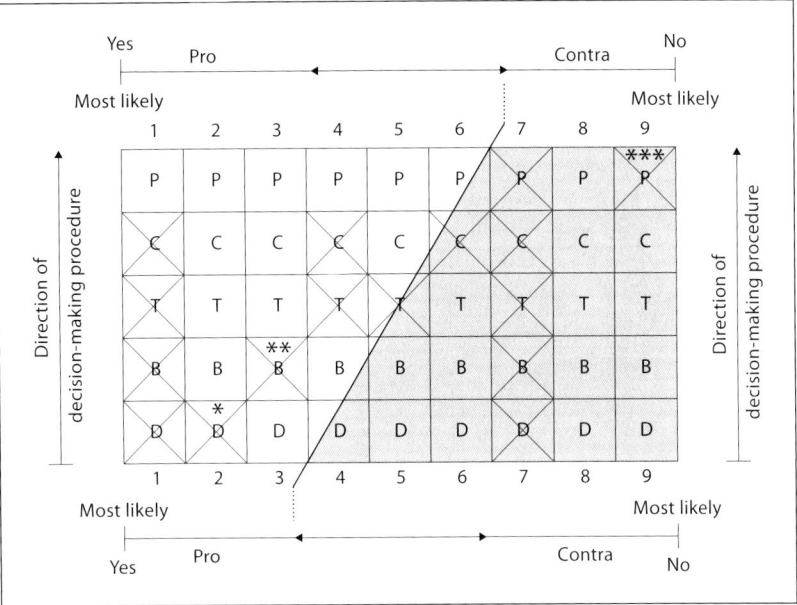

Fig. 1. The grid consists of nine columns. Each column consists of five domains. A crossed out domain means that characteristic/manifestation of this domain is non-favorable.

D, B, T, C, P = Influencing domains for decision of pro or contra FV supplementation. D = Diet analysis, B = biochemical analyses + clinical anamnesis, T = exercise training and competition, C = basic conditions, P = quality and scientific evidence of FV supplement/product.

1...9 = Columns with numbers. Each column consists of 5 influencing domains.

X = Characteristic/manifestation of domain is non-favorable for athlete's health and performance; in opposite to non-canceled squares (favorable characteristic/manifestation).

* = If diet (D) is non-favorable, FV supplementation is recommended. This is independent from all other domains (except for P). ** = If biochemistry (B) is non-favorable, FV supplementation is recommended. This is independent from all other domains (except for P). *** = If FV product (P) is non-favorable, FV supplementation with the certain product has to be avoided. This is independent from all other domains.

Columns right of diagonal dividing line = con FV supplementation; columns left of diagonal dividing line = pro FV supplementation.

Mixed right/left column No. 5 = pro or con depends on priorities and goals of exercise training/competition: (a) Priority is adaptation to exercise training: rather contra or fewer FV supplementation. (b) Priority is stabilization of immunity e.g. during longer lasting competition series: rather pro or increased FV supplementation.

Mixed right/left columns Nos. 4 and 6 = amount of colored area illustrates tendency to pro or con of supplementation with mixed FV concentrates. Dimension and degree of C and T are crucial for decision.

Concluding Remarks

The main base for each athlete's health, immunity and performance is a mixed balanced diet. Supplementation with mixed FVB concentrates is indicated if health stabilization and avoidance of oxidative stress and inflammation have priority. On the other hand, there is no real evidence that FVB supplements provide a direct beneficial impact on an athlete's performance.

It is not realistic to provide general advice pro or con regarding supplementation with mixed FVB concentrates. Pro or con and also amount and dosage of supplementation underlie individual evaluation of each influencing domain in every single athlete. The manifestation(s) of the five domains, diet, biochemical analyses and clinical history, exercise training, basic conditions and the product itself, are crucial for decision-making. Thus in practice, a counselor's decision pro or con FVB supplementation should be based on a systematic decision-making procedure. The 'domain grid' as provided in figure 1 in this article might guide to find the right solution.

In future, the number of exercise studies with these promising nutraceuticals should increase. The results from research in this field should be combined with practical observations and documentary reports to achieve sustainable health and performance of sporty people.

Disclosure Statement

The author has been awarded competitive research grants from NSA to study Juice Plus+®.

References

1 Ness AR, Powles JW: Fruit and vegetables, and cardiovascular disease: a review. Int J Epidemiol 1997; 26:1–13.

2 Dauchet L, Amouyel P, Hercberg S, et al: Fruit and vegetable consumption and risk of coronary heart disease: a meta-analysis of cohort studies. J Nutr 2006;136:2588–2593.

3 Nikolic M, Nikic D, Petrovic B: Fruit and vegetable intake and the risk for developing coronary heart disease. Cent Eur J Public Health 2008;16:17–20.

4 Igbal R, Anand S, Ounpuu S, et al: Dietary patterns and the risk of acute myocardial infarction in 52 countries: results of the INTERHEART study. Circulation 2008;118:1929–1937.

5 Pomerleau J, Lock K, McKee M: The burden of cardiovascular disease and cancer attributable to low fruit and vegetable intake in the European Union: differences between old and new member states. Public Health Nutr 2006;9:575–583.

6 Joshipura KJ, Ascherio A, Manson JE, et al: Fruit and vegetable intake in relation to risk of ischemic stroke. JAMA 1999;282:1233–1239.

7 Dauchet L, Amouyel P, Dallongeville J: Fruit and vegetable consumption and risk of stroke. Neurology 2005;65:1193–1197.

8 Wright ME, Park Y, Subar AF, et al: Intakes of fruit, vegetables, and specific botanical groups in relation to lung cancer risk in the NIH-AARP diet and health study. Am J Epidemiol 2008;168:1024–1034.

9 Nöthlings U, Schulze MB, Weikert C, et al: Intake of vegetables, legumes and fruit, and risk for all-cause, cardiovascular, and cancer mortality in a European diabetic population. J Nutr 2008;138:775–781.

10 Yamaji T, Inoue M, Sasazuki S, et al: Fruit and vegetable consumption and squamous cell carcinoma of the esophagus in Japan: the JPHC study. Int J Cancer 2008;123:1935–1940.

11 Steinmetz KA, Potter JD: Vegetable, fruit, and cancer prevention: a review. J Am Diet Assoc 1996;96: 1027–1039.

12 Bazzano LA, Li TY, Joshipura KJ, et al: Intake of fruit, vegetables, and fruit juices and risk of diabetes in women. Diabetes Care 2008;31:1311–1317.

13 Lampe JW: Health effects of vegetables and fruit: assessing mechanisms of action in human experimental studies. Am J Clin Nutr 1999;70:475S–490S.

14 Dragsted LO, Pedersen A, Hermetter A, et al: The 6-a-day study: effects of fruit and vegetables on markers of oxidative stress and antioxidative defense in healthy nonsmokers. Am J Clin Nutr 2004;79: 1060–1072.

15 Herrera E, Jimenez R, Aruoma OI, et al: Aspects of antioxidant foods and supplements in health and disease. Nutr Rev 2009;67:S140–S144.

16 Brown L, Rosner B, Willett WW, et al: Cholesterol-lowering effects of dietary fiber: a meta-analysis. Am J Clin Nutr 1999;69:30–42.

17 Casagrande SS, Wang Y, Anderson C, et al: Have Americans increased their fruit and vegetable intake? The trends between 1988 and 2002. Am J Prev Med 2007;32:257–263.

18 Naska A, Vasdekis VG, Trichopoulou A, et al: Fruit and vegetable availability among ten European countries: how does it compare with the five-a-day' recommendation? DAFNE I and II projects of the European Commission. Br J Nutr 2000;84:549–556.

19 Billson H, Pryer JA, Nichols R: Variation in fruit and vegetable consumption among adults in Britain. An analysis from the dietary and nutritional survey of British adults. Eur J Clin Nutr 1999;53:946–952.

20 German Ministry for Nutrition, Agriculture and Consumer Protection (ed): Nationale Verzehrsstudie II (NVS II), Ergebnisbericht Teil 2. Berlin, 2011.

21 Bundesamt für Gesundheit (ed): 5. Schweizerische Ernährungsbericht. Bern, 2005.

22 Elmadfa I, Freisling H, König J, et al: Austrian Nutrition Report 2008. Austrian Ministry of Health and Institute for Nutrition Sciences, University of Vienna (eds): Vienna, 2008.

23 WHO, Food and Agriculture Organization of the United Nations (ed): Diet, Nutrition and the Prevention of Chronic Disease. Tech Rep Ser 916. Geneva, WHO, 2003.

24 Danish Ministry of Food, Agriculture and Fisheries (ed): Fruits and vegetables – recommendations for intake. Copenhagen, 1998.

25 United States Department of Agriculture (ed): Dietary Guidelines for Americans 2010. Chapter 4: Foods and Nutrients to increase. Accessed January 28, 2012. http://www.cnpp.usda.gov/Publications/DietaryGuidelines/2010/PolicyDoc/Chapter4.pdf.

26 Ivy J, Portman R: The performance zone: your nutrition action plan for greater endurance and sports performance. Rosenberg C (ed): North Bergen, Basic Health Publ Inc, 2004.

27 Lamprecht M, Smekal G: Sport und Ernährung; in Pokan, Förster, Hofmann, Hörtnagl, Ledl-Kurkowski, Wonisch (eds): Kompendium der Sportmedizin. New York, Springer, 2004, pp 179–226.

28 Liu RH: Health benefits of fruit and vegetables are from additive and synergistic combinations of phytochemicals. Am J Clin Nutr 2003;78:517S–520S.

29 Oude Griep LM, Geleijnse JM, Kromhout D, et al: Raw and processed fruit and vegetable consumption and 10-year coronary heart disease incidence in a population-based cohort study in the Netherlands. PLoS One 2010;5:e13609.

30 Esfahani A, Wong JMW, Truan J, et al: Health effects of mixed fruit and vegetable concentrates: a systematic review of the clinical interventions. J Am Coll Nutr 2011;30:285–294.

31 Peternelj TT, Coombes JS: Antioxidant supplementation during exercise training. Sports Med 2011;41:1043–1069.

32 Chevion S, Moran DS, Heled Y, et al: Plasma antioxidant status and cell injury after severe physical exercise. Proc Natl Acad Sci USA 2003;100:5119–5123.

33 Petibois C, Deleris G: Evidence that erythrocytes are highly susceptible to exercise oxidative stress: FT-IR spectrometric studies at the molecular level. Cell Biol Int 2005;29:709–716.

34 Urso MI, Clarkson PM: Oxidative stress, exercise, and antioxidant supplementation. Toxicology 2003;189:41–54.

35 Saxton JM, Donnelly AE, Roper HP: Indices of free radical-mediated damage following maximum voluntary eccentric and concentric muscular work. Eur J Appl Physiol 1994;68:189–193.

36 Fischer CP, Hiscock NJ, Penkowa M, et al: Supplementation with vitamins C and E inhibits the release of interleukin-6 from contracting human skeletal muscle. J Physiol 2004;558:633–645.

37 Kuipers H: Exercise-induced muscle damage. Int J Sports Med 1994;15:132–135.

38 Peters EM, Goetzche JM, Grobbelaar B, et al: Vitamin C supplementation reduces the incidence of post-race symptoms of upper respiratory tract infection in ultradistance runners. Am J Clin Nutr 1993;57:170–174.

39 Casagrande SS, Gary-Webb TL: Trends in US adult fruit and vegetable consumption; in Watson RR, Preedy VR (eds): Bioactive Foods in Promoting Health, Fruits and Vegetables. San Diego, Academic Press, 2010, pp 111–130.

40 The European Food Information Council (EUFIC) (ed): EUFIC Review 01/2012. Accessed January 28, 2012. http://www.eufic.org/article/en/expid/Fruit-vegetable-consumption-Europe/

41 Austrian Ministry of Health (ed): Nationaler Aktionsplan Ernährung – NAPe, Vienna 2011.

42 German Ministry for Nutrition, Agriculture and Consumer Protection and Ministry of Health (eds): IN FORM – Deutschlands Initiative für gesunde Ernährung und mehr Bewegung, Berlin 2011.

43 Lamprecht M, Öttl K, Schwaberger G, et al: Supplementation with mixed fruit and vegetable juice concentrates attenuates oxidative stress markers in trained athletes. Med Sci Sports Exerc 2005;37:S46.

44 Bloomer RJ, Goldfarb AH, McKenzie MJ: Oxidative stress response to aerobic exercise: comparison of antioxidant supplements. Med Sci Sports Exerc 2006;38:1098–1105.

45 Goldfarb AH, McKenzie MJ, Bloomer RJ: Gender comparisons of exercise-induced oxidative stress: influence of antioxidant supplementation. Appl Physiol Nutr Metab 2007;32:1124–1131.

46 Lamprecht M, Oettl K, Schwaberger G, et al: Several indicators of oxidative stress, immunity and illness improved in trained men consuming an encapsulated juice powder concentrate for 28 weeks. J Nutr 2007;137:2737–2741.

47 Lamprecht M, Oettl K, Schwaberger G, et al: Protein modification responds to exercise intensity and antioxidant supplementation. Med Sci Sports Exerc 2009;41:155–163.

48 Lamprecht M, Hofmann P, Greilberger JF, et al: Increased lipid peroxidation in trained men after 2 weeks of antioxidant supplementation. Int J Sport Nutr Exerc Metab 2009;19:385–399.

49 Knez WL, Jenkins DG, Coombes JS: Oxidative stress in half and full Ironman triathletes. Med Sci Sports Exerc 2007;39:283–288.

50 Nieman DC, Henson DA, McAnulty SR, et al: Vitamin E and immunity after the Kona Triathlon World Championship. Med Sci Sports Exerc 2004;36:1328–1335.

51 Goldfarb AH, Garten RS, Cho C, et al: Effects of a fruit/berry/vegetable supplement on muscle function and oxidative stress. Med Sci Sports Exerc 2011; 43:501–508.

52 Silva L, Pinho C, Silveira P, et al: Vitamin E supplementation decreases muscular and oxidative damage but not inflammatory response induced by eccentric contraction. J Physiol Sci 2010;60:51–57.

53 Jakeman P, Maxwell S: Effect of antioxidant vitamin supplementation on muscle function after eccentric exercise. Eur J Appl Physiol Occup Physiol 1993;67: 426–430.

54 Palazzetti S, Rousseau AS, Richard MJ, et al: Antioxidant supplementation preserves antioxidant response in physical training and low antioxidant intake. Br J Nutr 2004;91:91–100.

55 Nakhostin-Roohi B, Babaei P, Rahmani-Nia F, et al: Effect of vitamin C supplementation on lipid peroxidation, muscle damage and inflammation after 30-min exercise at 75% $Vo_{2 max}$. J Sports Med Phys Fitness 2008;48:217–224.

56 Bloomer RJ, Goldfarb AH, McKenzie MJ, et al: Effects of antioxidant therapy in women exposed to eccentric exercise. Int J Sport Nutr Exerc Metab 2004;14:377–388.

57 Bryer SC, Goldfarb AH: Effect of high dose vitamin C supplementation on muscle soreness, damage, function, and oxidative stress to eccentric exercise. Int J Sport Nutr Exerc Metab 2006;16:270–280.

58 Teixeira VH, Valente HF, Casal SI, et al: Antioxidants do not prevent post-exercise peroxidation and may delay muscle recovery. Med Sci Sports Exerc 2009; 41:1752–1760.

59 Childs A, Jacobs C, Kaminski T, et al: Supplementation with vitamin C and N-acetyl-cysteine increases oxidative stress in humans after an acute muscle injury induced by eccentric exercise. Free Radic Biol Med 2001;31:745–753.

60 Close GL, Ashton T, Cable T, et al: Ascorbic acid supplementation does not attenuate post-exercise muscle soreness following muscle-damaging exercise but may delay the recovery process. Br J Nutr 2006;95:976–981.

61 Avery NG, Kaiser JL, Sharman MJ, et al: Effects of vitamin E supplementation on recovery from repeated bouts of resistance exercise. J Strength Cond Res 2003;17:801–809.

62 Weight LM, Myburgh KH, Noakes TD: Vitamin and mineral supplementation: effect on the running performance of trained athletes. Am J Clin Nutr 1988;47:192–195.

63 Nielsen AN, Mizuno M, Ratkevicius A, et al: No effect of antioxidant supplementation in triathletes on maximal oxygen uptake, ^{31}P-NMRS detected muscle energy metabolism and muscle fatigue. Int J Sports Med 1999;20:154–158.

64 Arent SM, Pellegrino JK, Williams CA, et al: Nutritional supplementation, performance, and oxidative stress in college soccer players. J Strength Cond Res 2010;24:1117–1124.

65 Zoppi CC, Hohl R, Silva FC, et al: Vitamin C and E supplementation effects in professional soccer players under regular training. J Int Soc Sports Nutr 2006;3:37–44.

66 Fry AC, Bloomer RJ, Falvo MJ, et al: Effect of a liquid multivitamin/mineral supplement on anaerobic exercise performance. Res Sports Med 2006;14: 53–64.

67 Knechtle B, Knechtle P, Schulze I, et al: Vitamins, minerals and race performance in ultra-endurance runners: Deutschlandlauf 2006. Asia Pac J Clin Nutr 2008;17:194–198.

68 Gomez-Cabrera MC, Domenech E, Romagnoli M, et al: Oral administration of vitamin C decreases muscle mitochondrial biogenesis and hampers training-induced adaptations in endurance performance. Am J Clin Nutr 2008;87:142–149.

69 Khassaf M, McArdle A, Esanu C, et al: Effect of vitamin C supplements on antioxidant defence and stress proteins in human lymphocytes and skeletal muscle. J Physiol 2003;549:645–652.

70 Fischer CP, Hiscock NJ, Basu S, et al: Vitamin E isoform-specific inhibition of the exercise-induced heat shock protein 72 expression in humans. J Appl Physiol 2006;100:1679–1687.

71 Gomez-Cabrera MC, Pallardo FV, Sastre J, et al: Allopurinol and markers of muscle damage among participants in the Tour de France. JAMA 2003;289: 2503–2504.

72 Mac Rae HS, Mefferd KM: Dietary antioxidant supplementation combined with quercetin improves cycling time trial performance. Int J Sport Nutr Exerc Metab 2006;16:405–419.

73 Bailey SJ, Winyard P, Vanhatalo A, et al: Dietary nitrate supplementation reduces the O_2 cost of low-intensity exercise and enhances tolerance to high-intensity exercise in humans. J Appl Physiol 2009; 107:1144–1155.

74 Skarpanska-Stejnborn A, Pilaczynska-Szczesniak L, Basta P, et al: The influence of supplementation with *Rhodiola rosea* L. extract on selected redox parameters in professional rowers. Int J Sport Nutr Exerc Metab 2009;19:186–199.

75 Vauzour D, Rodriguez-Mateos A, Corona G, et al: Polyphenols and human health: prevention of disease and mechanisms of action. Nutrients 2010;2: 1106–1131.

76 Roll S, Nocon M, Willich SN: Reduction of common cold symptoms by encapsulated juice powder concentrate of fruits and vegetables: a randomised, double-blind, placebo-controlled trial. Br J Nutr 2010;105:118–122.

77 Position of the American Dietetic Association: Health implications of dietary fiber. JADA 2002; 102:993–1000.

Manfred Lamprecht, PhD, PhD
Institute of Physiological Chemistry
Centre for Physiological Medicine, Medical University of Graz
Harrachgasse 21/II, Graz (Austria)
Tel. +43 664 1555528, E-Mail manfred.lamprecht@medunigraz.at, and
Green Beat – Institute of Nutrient Research and Sport Nutrition
Petersbergenstrasse 95b, AT-8042 Graz (Austria)
E-Mail office@greenbeat.at

Lamprecht M (ed): Acute Topics in Sport Nutrition.
Med Sport Sci. Basel, Karger, 2013, vol 59, pp 86–93

Cherry Juice Targets Antioxidant Potential and Pain Relief

Kerry S. Kuehl

Division of Health Promotion and Sports Medicine, Department of Medicine, Oregon Health and Science University, Portland, Oreg., USA

Abstract

Strenuous physical activity increases the risk of musculoskeletal injury and can induce muscle damage resulting in acute inflammation and decreased performance. The human body's natural response to injury results in inflammation-induced pain, swelling, and erythema. Among sports medicine physicians and athletic trainers, the mainstays of urgent treatment of soft tissue injury are rest, ice, compression, and elevation (RICE). In order to reduce pain and inflammation, anti-inflammatory agents such as non-steroidal anti-inflammatory drugs (NSAIDs) act on the multiple inflammatory pathways, which, although often very effective, can have undesirable side effects such as gastric ulceration and, infrequently, myocardial infarction and stroke. For centuries, natural anti-inflammatory compounds have been used to mediate the inflammatory process and often with fewer side effects. Tart cherries appear to possess similar effectiveness in treating the inflammatory reaction seen in both acute and chronic pain syndromes encountered among athletes and non-athletes with chronic inflammatory disease. This article reviews the antioxidant and anti-inflammatory effects of tart cherries on prevention, treatment, and recovery of soft tissue injury and pain.

Epidemiological evidence suggests that a high intake of plant foods is associated with lower risk of chronic diseases, and specifically, consumption of fruit and vegetables is associated with decreased risk of cancer and heart disease [1, 2]. Research into the possible mechanism has identified numerous antioxidant and anti-inflammatory agents in plants purported to reduce illness and disease associated with inflammation and tissue damage. These disease-modifying agents include cyclooxygenase inhibitory flavonoids [3, 4] and anthocynanins with high antioxidant and anti-inflammatory activities that have been identified in foods from black tea to tart cherries to fish oil [5–8]. Most recently, there has been interest in the role of these potent natural phytochemicals to reduce musculoskeletal injury, inflammation and pain, and improve recovery from exercise-induced soft tissue muscle damage [9–12]. Intense physical activity increases the risk of musculoskeletal injury and can induce acute muscle damage resulting in acute inflammation and decreased force production [13–17]. The mainstays of treatment of soft tissue injury and inflammation are rest, ice, compression, elevation (RICE) and non-steroidal anti-inflammatory drugs (NSAIDs) [18, 19]. According to the US Food and Drug Administration in 2001, NSAIDs accounted for 70,000,000 prescriptions and 30 billion over-the-counter doses sold annually in the USA, making it the most used and prescribed medication in the USA. There are, however, known adverse effects associated with the use of NSAIDs [20] and natural anti-inflammatory agents may be a beneficial and safer alternative [21]. Consumption of cherries may be effective in prevention of oxidative damage causing disease processes and alleviating symptoms in inflammatory conditions [9, 12, 22–24]. Consumption of approximately 45 cherries per day has been shown to reduce circulating concentrations of inflammatory markers in healthy men and women [25].

Mechanism of Inflammation and Traditional Pharmacologic Therapy

Strenuous physical activity increases the risk of musculoskeletal injury and can induce acute muscle damage resulting in acute inflammation and decreased force production that can last up to 1 week post-exercise [16, 17, 26–30]. The exact mechanism by which muscle damage occurs is not completely understood, but processes involve both mechanical and metabolic pathways. Pain, heat, redness, and swelling (dolor, calor, rubor, tumor) are the classic manifestations of the inflammatory process. The acute phase of muscle damage is caused by extensive myofibril disruption triggering a localized inflammatory response including release of proinflammatory cytokines interleukin (IL)-1α, IL-1β, IL-6 and tumor necrosis factor (TNF-α). Increased concentrations of TNF-α are believed to cause the cardinal signs of inflammation to occur. In addition, these cytokines increase production of leukotrienes leading to increased vascular permeability, attracting neutrophils to the injury site, and resulting in free radical production [31, 32]. When these muscle fibers are exposed to this oxidative stress due to exercise-induced increases in reactive oxygen species (ROS) and nitric oxide (NO) derivatives that exceed the antioxidant defense capacity [13, 14, 33]. With the elucidation of the role of these inflammatory cytokines and the apparent role for ROS and NO in muscle damage, there has been considerable interest in the efficacy of antioxidant supplements in ameliorating exercise-induced muscle damage, and foods that can alleviate inflammation and relieve pain [34].

Antioxidant Capacity of Tart Cherries

Both sweet and tart cherries are rich in phenolic compounds, but tart cherries are considered to be one of the highest sources of phenolic compounds, including cyclooxygenase inhibitory flavonoids and anthocyanins, with high levels of antioxidant and anti-inflammatory activity [3–5, 8]. The levels of anthocyanins and other flavonoids in the Montmorency and Balaton tart cherry were analyzed comparatively by high-performance liquid chromatography and electrospray mass spectroscopy showing the major anthocyanin compound in both of these cultivars is cyanidin-3-glucosylrutinoside, followed by cyanidin-3-rutinoside and peonidin-3-glucoside [35]. Studies on the antioxidant activities (total antioxidant status assay) of crude tart cherry extracts including juice show that these products preserve their antioxidant capacities after processing and storage [36]. When the trolox equivalent antioxidant capacity values were evaluated conceptually against the cherry phytochemical profile, cyanidin and its derivatives were found to be significant contributors to the antioxidant systems of tart cherries. One of the best known properties of anthocyanidins is their strong antioxidant activity in metabolic reactions, due to their ability to scavenge oxygen free radicals and other ROS. This biological feature makes the tart cherry a significant antioxidant food source impacting oxidative damage processes. In animal studies, tart cherry-enriched diets reduced oxidative stress and inflammation [37] as well as protective effects on neuronal cells [38] and inhibition of tumorigenesis [39].

Tart Cherry Flavonoids as Anti-Inflammatory Agents

The anti-inflammatory actions of flavonoids in vitro or in cellular models involve the inhibition of the synthesis and activities of different proinflammatory mediators such as eicosanoids, cytokines, adhesion molecules and CRP [5]. Acute inflammation as stated above is an essential and complex response protecting the body against harmful stimuli such as pathogens, damaged cells, or other irritants, and is manifested by vascular changes, edema, and predominantly neutrophilic infiltration in a matter of days. Chronic inflammation is characterized by prolonged duration (weeks or months) caused by persistent infections, immune-mediated inflammatory diseases, or prolonged exposure to toxic reagents. This results in severe tissue destruction caused predominantly by mononuclear macrophages. Macrophages are the dominant cellular player in chronic inflammation, with a lifespan of several months to years [40]. In a study looking at the anti-inflammatory mechanism of sweet and sour cherries, it was demonstrated that cherries have an anti-inflammatory effect mechanism similar to traditional NSAIDs. Water extracts of Balaton and Montmorency tart cherries inhibited cyclooxygenase-1 and cyclooxygenase-2 enzyme's effect by 84, 91, 77, and 87% respectively, at 250 μg/ml per dose [40].

Both sweet and tart cherries are rich in phenolic compounds, but tart cherries are considered to be one of the highest sources of phenolic compounds, including cyclooxygenase inhibitory flavonoids and anthocyanins, with high levels of antioxidant and anti-inflammatory activity [3–5, 8, 40]. This has led to speculation that cherry consumption may be beneficial among patients with chronic pain and inflammation disease processes like arthritis, gout, and fibromyalgia. Cherry consumption relieved symptoms of arthritis in an early human trial [41] and more recently, a single bolus of Bing cherries administered to healthy women after a 12-hour fast showed a trend for reducing circulating concentrations of CRP and NO, and uric acid within 3 h of the bolus, but not statistically significant ($p < 0.1$) [23]. In a follow-up study by Kelley et al. [25] in 2006, consumption of approximately 45 cherries per day was shown to reduce circulating concentrations of inflammatory markers in healthy men and women.

Tart cherry juice is purported to benefit patients with fibromyalgia, perhaps due to its antioxidant scavenging and anti-inflammatory properties [42]. Fibromyalgia is a common chronic pain disorder, and physical activity is often used to manage this illness [43]. However, exercise can cause heightened discomfort following exertion, and tart cherry juice may alleviate the symptoms of delayed-onset muscle soreness associated with exercise among fibromyalgia patients. To test, the efficacy of tart cherry juice to maintain strength and reduce pain undergoing arm exercise, 14 female subjects with fibromyalgia ingested tart cherry juice or placebo for 10 days in a blinded, randomized, crossover design [11]. Ingestion of tart cherry juice demonstrated marginal benefits in maintaining muscle strength 24 h after strenuous exercise. In addition, a subset of participants who ingested the tart cherry juice had a significant reduction in pain after the eccentric arm exercise stress. Finding one subset of subjects who benefitted more than others was provocative. The response to most therapies for fibromyalgia is variable [42] and unlike other studies, this study was blinded in a crossover design adding validity to the observed effects of a substantial reduction in pain in the tart cherry group as compared to placebo.

Osteoarthritis (OA) is a common syndrome affecting 65 million Americans characterized by pain and disability [44, 45]. Pain relief and improvement of functional disability are the main goals of treatment. Over 40% of OA has an inflammatory component and the standardization of therapeutic criteria for inflammatory OA has stimulated much research, and understanding of the wide variety of therapeutic approaches is complicated by anecdotal and non-evidence-based basis of OA [45, 46]. Non-pharmacologic interventions are the mainstay of treatment (patient education, exercise, occupational therapy), but oral medications and nutritional supplements have shown positive and negative results, with acetaminophen being the most common pain medication [47, 48]. A number of studies have looked at dietary factors and some have improved arthritis pain and function [49]. A recent study looking at the effects of tart cherry on inflammatory biomarkers among inflammatory OA patients showed a beneficial effect on serum inflammation indices [50]. In a double-blind, randomized placebo-controlled trial, 20 female subjects between 50 and 70 years diagnosed with Inflammatory OA consumed either placebo or Montmorency tart cherry juice (two 10-oz bottles of juice was equivalent to about 100 tart cherries per day) for 3 weeks. Among subjects consuming the tart cherry juice as compared to the placebo beverage, a statistically significant decrease in CRP was observed. This study showed the value of looking at alternative therapies to conventional methods in the treatment and management of chronic inflammatory conditions. Tart cherry juice may provide beneficial anti-inflammatory activity helping OA patients manage their disease with less adverse effects than traditional arthritis medications.

Effects of Tart Cherries on Muscle Injury and Recovery

Muscle damage, inflammation, and oxidative stress typically occur in response to high intensity or prolonged physical activity [13, 14, 22, 23]. Numerous studies have examined the effect of dietary supplements that contain antioxidants on muscle function, performance, and on markers of muscle damage and inflammation with mixed results. Kuehl et al. [12] studied the effects of tart cherry juice consumption versus placebo in a double-blinded, randomized trial design of runners participating in a 24-hour relay race in Oregon. It is well documented that running for distances in excess of typical training distances causes acute muscle injury, and that eccentric muscle actions, such as downhill running, exacerbate injury and soreness. The Hood to Coast relay requires participants to run three separate race segments over an approximately 24-hour period, including segments that ascend or descend steep terrain from Mt Hood Oregon (started at 5,500 ft elevation) to Seaside, Oregon (sea level) traversing the Cascade and Coastal mountain ranges. Runners drank either two 10-oz tart cherry or placebo beverage daily for 1 week prior to the race and during the race. Both groups (tart cherry and placebo) reported significantly higher pain levels based on the visual analog scale (VAS) upon completion of the race. However, participants who drank the tart cherry juice twice daily for 1 week prior to and the day of the race reported a significantly smaller increase in pain after the race (mean post-race VAS score was 12 mm in tart cherry juice as compared to a VAS score of 37 mm). The 25-mm difference in the mean VAS score is equivalent to a therapeutic pain relief dose of ibuprofen among runners consuming the tart cherry juice [19]. These results suggest tart cherries provide a protective benefit against the acute muscle soreness caused by distance running.

Tart cherry juice may prevent symptoms of muscle damage among individuals participating in strenuous exercise. Looking at the effects of tart cherry juice on indices of recovery following a long-distance running event, Howatson et al. [22] evaluated marathon runners who drank cherry juice 5 days before, the day of, and after the race. Results showed those who drank cherry juice had statistically significant lower levels of inflammation markers (IL-6 and CRP) as compared to placebo. In addition to biomarkers, isometric strength recovered significantly faster in the tart cherry juice runners indicating a viable means to aid recovery following strenuous exercise. A study among healthy, exercise-naive individuals demonstrated

women [6], for those wanting to lose weight [7] and for the elderly [8, 9]. Maintaining or increasing muscle mass and reducing fat mass is an important priority for athletes and those conscious of body image, but also from a health standpoint, it is also of pivotal importance for preventing and treating obesity and other chronic cardiometabolic and musculoskeletal diseases including diabetes, heart disease and osteoporosis [10–13].

Milk contains a variety of bioactive components, including protein, calcium, and vitamin D, that may act independently or synergistically to improve body composition during periods of energy balance [6] or energy restriction [7, 14, 15]. In addition, milk consumption following resistance exercise has been shown to increase muscle protein synthesis and lean mass to a greater extent than other similar protein sources (i.e. soy) [5, 16]. Thus, given the composition of milk/dairy foods and its physiological effects, it would seem to be an excellent post-resistance exercise food to help augment positive body composition change not only in athletes, but also in recreationally active and newly active persons or those seeking to lose body weight as fat.

Resistance training is not as common an exercise modality for women as it is for men. Reasons for this may be that women generally choose to avoid exercise that they perceive might contribute to muscle 'bulk', potential weight gain or less weight loss [17]. Moreover, milk and other dairy products are often avoided by women since they are cited as being fattening [18]. Thus, only a handful of studies have been carried out assessing the effect of resistance exercise and either milk, yogurt or other dairy products on body composition in women. In addition, no such studies have been undertaken in female athletes. Hence, this chapter highlights results from clinical studies carried out in women undergoing acute resistance exercise or chronic resistance training with the consumption of dairy, and explores the effect of this combined paradigm on body composition, namely fat, lean (muscle) and bone mass.

Milk and Body Composition – Mechanism of Action

Milk contains calcium (300 mg/cup), vitamin D (90 IU/cup) and protein (9 g/cup), each of which has been shown to modulate body composition [2, 3, 19, 20]. Calcium and vitamin D affect adipose tissue metabolism (stimulating lipolysis and inhibiting fat storage) [20, 21], improve bone health (inhibiting osteoclast resorption of bone) [10, 22], and decrease intestinal fat absorption by binding to fatty acids and forming insoluble soaps and by destabilizing the formation of micelles in the gastrointestinal tract [23].

Insofar as protein is concerned, a large part of dairy's bioactivity is in the whey protein fraction which is rich in branched chain amino acids (BCAA) [24]. In fact, BCAAs make up about 25% of total dairy proteins [25]. Leucine, a BCAA found in relatively high abundance in whey protein, has been shown to affect both fat and muscle metabolism [26, 27]. In vitro, leucine inhibits triglyceride synthesis (lipogenesis), promotes fat cell breakdown (lipolysis), and increases fat oxidation in muscle cells [27]. In addition, and perhaps as its better known role, leucine stimulates the translation-initiation machinery for muscle protein synthesis [26]. The particulars of this mechanism will not be discussed here, but the effect of leucine on muscle protein synthesis may be one of the key ways in which milk and dairy foods augment muscle mass accretion following resistance exercise [1–3, 26], and promote the sparing of muscle during weight loss [7, 15].

Milk, Whey Protein and Resistance Exercise – Acute Response

Two recent acute studies assessed the effect of milk or dairy-derived proteins on muscle protein synthesis in women [28, 29]. Both of these studies employed a resistance exercise bout, *vastus lateralis* muscle biopsies and stable isotope tracer methodologies, however, neither were carried out exclusively in women. The first study assessed the acute ingestion of either fat-free milk or whole milk 60 min after resistance exercise. Both types of milk stimulated greater net uptake of amino acids representing increased rates of muscle protein synthesis following resistance exercise and milk consumption [28]. However, whole milk stimulated significantly greater threonine uptake suggesting that high-fat milk may have increased utilization of available amino acids for muscle accretion. This is an interesting observation given that the only relevant difference between whole milk and fat-free milk that could impact protein metabolism is the fat content. To our knowledge, there are no data in humans to suggest that added fat improves muscle protein synthesis. Added fat adds more energy to the 'system', which has been shown to minimally impact the protein synthetic response [30]. Nevertheless, since the difference between milk types is minimal, recommendations to consume whole fat milk over fat-free milk for improved muscle mass accretion are without good basis and require replication before they can be made.

The second study by West et al. [29] was designed to uncover sex-based differences in muscle protein synthesis following resistance exercise and ingestion of 25 g of whey protein. No differences in rates of muscle protein synthesis were observed between the sexes. Therefore, both studies suggest that the combination of milk or its whey protein component and resistance exercise can acutely increase the rate of muscle protein accretion in women. However, neither study compared milk or whey to a post-exercise control beverage devoid of protein or to a different protein source; this is clearly an area for future study.

Dairy and Resistance Training – Chronic Studies

Given that milk and whey robustly increase muscle protein synthesis in the acute setting, it is important to investigate whether this effect translates into longer-term phenotypic changes, i.e. greater lean (muscle) mass gains. Five studies, four carried out in women only and one in a mixed sample, under conditions of weight maintenance or mild energy deficit (–250 kcal/day [31]), combined resistance exercise with milk, yogurt, whey protein, or a milk-like supplement for durations ranging from 6 to 24 weeks [6, 31–34].

As previously mentioned, whey protein provides all of the essential amino acids that are required for muscle protein synthesis [3, 26]. Soy protein is also considered a high-quality protein (i.e. Protein Digestibility Corrected Amino Acid Score (PDCAAS) = 1.0), however several studies [5, 16] have shown milk/whey to be superior, or at least equivalent [32, 35], to soy for skeletal muscle mass accretion. The duration of the training study may also be of importance. Studies shorter than 8 weeks in duration (training ~3×/week), usually considered the minimal amount of time to detect hypertrophy, may not be enough to observe significant differences in lean mass between treatments with different protein sources. Nonetheless, after just 6 weeks, trends were apparent in a study by Candow et al. [32] (two thirds were female participants), with lean mass gains of 2.5 and 1.7 kg in the whey vs. soy groups, respectively, and both protein groups showed significantly greater lean mass gains versus placebo (0.3 kg).

Yogurt contains most of the necessary bioactive components (except vitamin D unless fortified) to promote positive body composition change with resistance exercise. Due to its generally pleasant taste, and ease of consumption and digestion, it is an attractive way to provide the body with high-quality dairy post-resistance exercise. Despite this, only two studies of 8 [34] and 16 [31] weeks' duration have utilized yogurt (6 oz) as a post-exercise supplement, and both failed to demonstrate greater increases in lean mass compared to an isoenergetic carbohydrate-containing control. However, the yogurt supplements in both studies contained only 5 g of dairy protein, which is an amount that is substantially less than what is thought to be required (i.e. 20 g) to maximally stimulate muscle protein synthesis post-resistance exercise [36]. Thus, the lack of adequate amino acids during the critical post-exercise anabolic window [1] was probably responsible for a substandard anabolic stimulus for muscle protein synthesis and the lack of a superior gain in muscle protein mass. Nonetheless, we propose that yogurt would be an appropriate post-exercise dairy food if the right amount of protein is consumed. Future studies using yogurts with a higher per-serving protein content (i.e. some 'Greek-style' yogurts contain 16 g protein per 6 oz) may provide greater benefit from a body composition standpoint.

Only one resistance training study assessed body composition in young women after comparing an isoenergetic carbohydrate beverage to fat-free milk [6]. After 12 weeks, results showed that consuming 1 litre of milk post-resistance exercise (500 ml immediately after and an additional 500 ml 1 h after performing resistance exercise) resulted in greater lean mass gains, but also promoted a significantly greater loss of fat mass. The findings of this study are particularly relevant in terms of the beneficial change in body composition that occurred, i.e. concomitant fat mass loss and lean

Fig. 1. Change from baseline (mean ± SE) in lean mass (positive y-axis of graph) and fat mass (negative y-axis of graph) after young normal weight women underwent 12 weeks of resistance training with either milk or carbohydrate drink consumption post-exercise in Josse et al. [6]. * Significantly different from baseline, p < 0.05. † Significantly different from CON in same tissue, p < 0.05.

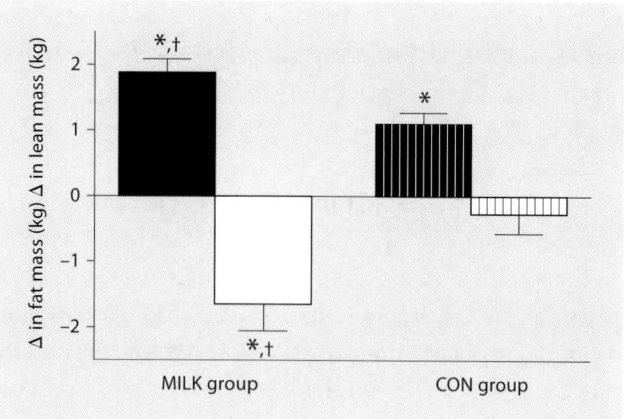

mass gain, in the face of very little (non-significant) change in body weight (fig. 1). On the other hand, the women consuming the isoenergetic carbohydrate beverage also showed lean mass gains, albeit significantly less than the milk group, and no loss of fat mass. The additional milk consumption-induced fat mass loss observed may be the cumulative result of the previously mentioned lipolytic mechanisms relating to other bioactive components in milk such as leucine, calcium and vitamin D. One other 24-week resistance training study carried out in early post-menopausal women utilized a daily supplement that resembled milk versus a placebo supplement [33]. The milk-like supplement contained whey protein (10 g), calcium (250 mg), vitamin D (200 IU), fat (1 g) and carbohydrate (31 g). They observed a more marked improvement in lean mass and strength in the milk-like group compared to the placebo [33].

Dairy, Resistance Training and Bone Health – Chronic Studies

Calcium and protein, two major components of milk, are essential for bone health [19]. They affect the structural integrity and strength of bone by influencing bone mineralization and collagen formation, respectively [10, 37]. Moreover, higher intakes of dietary calcium and vitamin D reduce circulating parathyroid hormone concentrations, which results in a positive effect on bone mineral density (BMD) and reduced rates of bone turnover [10]. Three studies have examined the effect of resistance exercise and dairy on bone health outcomes in women [6, 33, 38]; one study was carried out under slight energy restriction [38]. Compared to placebo treatments, lumbar spine BMD improved with resistance exercise and a high dairy-based calcium diet after 16 weeks in young women [38], and femoral neck BMD (adjusted for covariates) improved after consumption of a milk-like supplement for 24 weeks in early post-menopausal women [33]. In the third study, resistance training and milk consumption tended to increase bone formation (osteocalcin) and decrease resorption (C-telopeptide (CTx)), although differences between groups were not significant [6].

Milk/Dairy, Resistance Training and Weight Loss

Although no studies have directly examined the effect of exclusive resistance training and dairy during energy restriction (−500 kcal/day) on body composition, two clinical trials implementing a structured resistance training program along with aerobic exercise have. These studies were carried out in overweight or obese young women [7, 39, 40]. In the first study, after 12 weeks of energy restriction and exercise, no differences were observed between the four treatment groups (calcium lactate (1,500 mg/day), calcium phosphate (1,500 mg/day), milk/dairy (3 servings/day), placebo (baseline diet containing 750 mg of calcium)) for weight loss, fat loss or bone formation. In fact, the milk/dairy group lost the least amount of body fat compared to the placebo group [40]. While total energy intakes were not significantly different between the groups, the milk/dairy group consistently consumed >100 kcal/day more than the placebo group and no information was presented on macronutrient intakes. Therefore, it is possible that those in the milk/dairy group did not completely compensate for the added milk/dairy calories, and in fact, had an increased total energy intake. In addition, a less sensitive method (bioelectric impedance analysis) was used to assess fat mass, and this may also explain, at least in part, the lack of greater fat mass loss in this group [40].

In contrast to the findings seen in the other study [40], a combination of lifestyle (diet and exercise) factors and good support for weight loss, including increased intakes of dairy (milk, cheese and yogurt) and dietary protein resulted in a highly favourable body composition change characterized by greater total and visceral fat loss, increased lean mass and improved bone health versus diets lower in protein and low in dairy [7, 39]. The *Improving Diet, Exercise and Lifestyle (IDEAL) for Women* study was a 16-week clinical trial that employed a 2×/week structured resistance training program (split upper and lower body routine), in addition to required daily aerobic exercise. Based on previous weight loss research highlighting the effectiveness of dairy [14, 15], calcium [41], higher protein diets [42], and exercise [42, 43] in promoting positive body composition change in young women, we combined all of these factors, for the first time, on the background of energy restriction. A major aim of the *IDEAL for Women* study was to have subjects achieve weight loss of the highest possible 'quality', i.e. preferential loss of fat mass and preservation of lean mass and bone health. Figure 2 displays the fat and lean mass results from the *IDEAL for Women* study. Although total body weight loss was the same across all three groups, the composition of this loss was markedly different [7]. It is clear that the manipulations undergone by the high-protein-high-dairy (HPHD: 30% kcal/day protein,

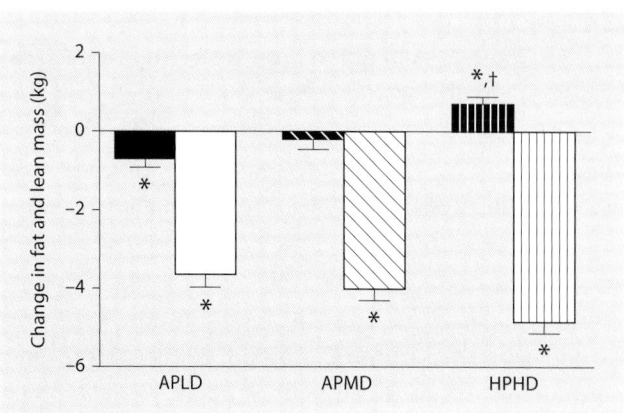

Fig. 2. Change from baseline (mean ± SE) in lean mass (black and textured bars) and fat mass (white and textured bars) after overweight and obese women underwent 16 weeks of diet and exercise-induced weight loss in the *IDEAL for Women* study [7]. APLD = Adequate-protein-low-dairy; APMD = adequate-protein-medium-dairy; HPHD = high-protein-high-dairy. * Significantly different from baseline, p < 0.05. † Significantly different from APLD, p < 0.05 (lean mass).

15% (i.e. half of the total dietary protein) as dairy protein, ~1,800 mg/day calcium) group succeeded in achieving high-quality weight loss. Furthermore, even in the face of energy restriction, which is a fundamentally catabolic process [44], these women continued to build lean tissue. The adequate-protein-medium-dairy (APMD: 15% kcal/day protein, 8% as dairy protein, ~1,200 mg/day calcium) group maintained lean mass with almost all of their weight loss as body fat, and this too represents a positive body composition change. Insofar as bone health is concerned, the two groups consuming dairy foods (HPHD and APMD) showed net increases in bone collagen formation (procollagen 1 amino-terminal propeptide [P1NP]/CTx ratio) and decreases in serum parathyroid hormone [39], whereas the opposite was observed for the adequate-protein-low-dairy (APLD: 15% kcal/day protein, 0% as dairy protein, ~300 mg/day calcium) group.

Conclusion

Milk and other dairy foods contain bioactive components (protein (leucine), calcium, vitamin D) that, in combination with resistance exercise [6, 7, 33, 38, 39], have been shown to consistently improve body composition in women under both iso- and hypoenergetic conditions. Acute studies demonstrate that resistance exercise with milk of different kinds [28] or whey [29] stimulate muscle protein synthesis in women, and that the myofibrillar synthetic response is no different from that observed in men [29]. Chronic resistance training studies confirm that acute muscle

Josse · Phillips

tissue accrual indeed translates into longer-term lean mass gains in women, but only when dairy protein is consumed at a sufficient quantity (≥15 g protein post-exercise [36]). In addition, concomitant reductions in fat mass were apparent under isoenergetic conditions [6], and bone health following resistance training with increased dairy/calcium intakes was improved [6, 33, 38]. With weight loss, the *IDEAL for Women* study highlighted the considerable benefit of consuming hypoenergetic diets higher in dairy (and calcium) and dietary protein with aerobic and resistance exercise on body composition (fat, lean and bone mass) in otherwise healthy overweight and obese young women [7, 39].

Overall, resistance exercise and milk/dairy consumption positively impacts body composition in women by promoting favourable changes in all three body compartments: fat, lean (muscle) and bone. Future research is required to assess the effect of milk and dairy with resistance training on body composition in female athletes. This, along with previously published work, will not only help to further promote milk and dairy for exercise recovery, but also help increase its consumption adding high-quality protein, calcium and vitamin D to the diets of young women.

Disclosure Statement

The authors have no conflicts of interest to disclose.

References

1 Burd NA, Tang JE, Moore DR, Phillips SM: Exercise training and protein metabolism: influences of contraction, protein intake, and sex-based differences. J Appl Physiol 2009;106:1692–1701.

2 Tang JE, Phillips SM: Maximizing muscle protein anabolism: the role of protein quality. Curr Opin Clin Nutr Metab Care 2009;12:66–71.

3 Phillips SM, Hartman JW, Wilkinson SB: Dietary protein to support anabolism with resistance exercise in young men. J Am Coll Nutr 2005;24:134S–139S.

4 Biolo G, Tipton KD, Klein S, Wolfe RR: An abundant supply of amino acids enhances the metabolic effect of exercise on muscle protein. Am J Physiol 1997;273:E122–E129.

5 Hartman JW, Tang JE, Wilkinson SB, Tarnopolsky MA, Lawrence RL, Fullerton AV, Phillips SM: Consumption of fat-free fluid milk after resistance exercise promotes greater lean mass accretion than does consumption of soy or carbohydrate in young, novice, male weightlifters. Am J Clin Nutr 2007;86:373–381.

6 Josse AR, Tang JE, Tarnopolsky MA, Phillips SM: Body composition and strength changes in women with milk and resistance exercise. Med Sci Sports Exerc 2010;42:1122–1130.

7 Josse AR, Atkinson SA, Tarnopolsky MA, Phillips SM: Increased consumption of dairy foods and protein during diet- and exercise-induced weight loss promotes fat mass loss and lean mass gain in overweight and obese premenopausal women. J Nutr 2011;141:1626–1634.

8 Esmarck B, Andersen JL, Olsen S, Richter EA, Mizuno M, Kjaer M: Timing of postexercise protein intake is important for muscle hypertrophy with resistance training in elderly humans. J Physiol 2001;535:301–311.

9 Koopman R, Verdijk L, Manders RJ, Gijsen AP, Gorselink M, Pijpers E, Wagenmakers AJ, van Loon LJ: Co-ingestion of protein and leucine stimulates muscle protein synthesis rates to the same extent in young and elderly lean men. Am J Clin Nutr 2006;84:623–632.

10 Heaney RP: Dairy and bone health. J Am Coll Nutr 2009;28(suppl 1):82S–90S.

11 Perez-Martin A, Raynaud E, Mercier J: Insulin resistance and associated metabolic abnormalities in muscle: effects of exercise. Obes Rev 2001;2:47–59.

12 Weinheimer EM, Sands LP, Campbell WW: A systematic review of the separate and combined effects of energy restriction and exercise on fat-free mass in middle-aged and older adults: implications for sarcopenic obesity. Nutr Rev 2010;68:375–388.

13 Wolfe RR: The underappreciated role of muscle in health and disease. Am J Clin Nutr 2006;84:475–482.

14 Zemel MB, Richards J, Mathis S, Milstead A, Gebhardt L, Silva E: Dairy augmentation of total and central fat loss in obese subjects. Int J Obes (Lond) 2005;29:391–397.

15 Zemel MB, Richards J, Milstead A, Campbell P: Effects of calcium and dairy on body composition and weight loss in African-American adults. Obes Res 2005;13:1218–1225.

16 Wilkinson SB, Tarnopolsky MA, Macdonald MJ, Macdonald JR, Armstrong D, Phillips SM: Consumption of fluid skim milk promotes greater muscle protein accretion after resistance exercise than does consumption of an isonitrogenous and isoenergetic soy-protein beverage. Am J Clin Nutr 2007;85:1031–1040.

17 Prichard I, Tiggemann M: Relations among exercise type, self-objectification, and body image in the fitness centre environment: the role of reasons for exercise. Psychol Sport Exer 2008;9:855–866.

18 Gulliver P, Horwath CC: Assessing women's perceived benefits, barriers, and stage of change for meeting milk product consumption recommendations. J Am Diet Assoc 2001;101:1354–1357.

19 Heaney RP, Layman DK: Amount and type of protein influences bone health. Am J Clin Nutr 2008;87:1567S–1570S.

20 Zemel MB: Mechanisms of dairy modulation of adiposity. J Nutr 2003;133:252S–256S.

21 Shi H, Norman AW, Okamura WH, Sen A, Zemel MB: 1α,25-Dihydroxyvitamin D_3 modulates human adipocyte metabolism via nongenomic action. FASEB J 2001;15:2751–2753.

22 Weaver CM, Heaney RP: Dairy consumption and bone health. Am J Clin Nutr 2001;73:660–661.

23 Christensen R, Lorenzen JK, Svith CR, Bartels EM, Melanson EL, Saris WH, Tremblay A, Astrup A: Effect of calcium from dairy and dietary supplements on faecal fat excretion: a meta-analysis of randomized controlled trials. Obes Rev 2009;10:475–486.

24 Blomstrand E, Eliasson J, Karlsson HK, Kohnke R: Branched-chain amino acids activate key enzymes in protein synthesis after physical exercise. J Nutr 2006;136(suppl):269S–273S.

25 Bos C, Gaudichon C, Tome D: Nutritional and physiological criteria in the assessment of milk protein quality for humans. J Am Coll Nutr 2000; 19(suppl):191S–205S.

26 Layman DK: Role of leucine in protein metabolism during exercise and recovery. Can J Appl Physiol 2002;27:646–663.

27 Sun X, Zemel MB: Leucine and calcium regulate fat metabolism and energy partitioning in murine adipocytes and muscle cells. Lipids 2007;42:297–305.

28 Elliot TA, Cree MG, Sanford AP, Wolfe RR, Tipton KD: Milk ingestion stimulates net muscle protein synthesis following resistance exercise. Med Sci Sports Exerc 2006;38:667–674.

29 West DW, Burd NA, Churchward-Venne TA, Camera DM, Mitchell CJ, Baker SK, Hawley JA, Coffey VG, Phillips SM: Sex-based comparisons of myofibrillar protein synthesis after resistance exercise in the fed state. J Appl Physiol 2012;112:1805–1813.

30 Svanberg E, Moller-Loswick AC, Matthews DE, Korner U, Andersson M, Lundholm K: The role of glucose, long-chain triglycerides and amino acids for promotion of amino acid balance across peripheral tissues in man. Clin Physiol 1999;19:311–320.

31 Thomas DT, Wideman L, Lovelady CA: Effects of a dairy supplement and resistance training on lean mass and insulin-like growth factor in women. Int J Sport Nutr Exerc Metab 2011;21:181–188.

32 Candow DG, Burke NC, Smith-Palmer T, Burke DG: Effect of whey and soy protein supplementation combined with resistance training in young adults. Int J Sport Nutr Exerc Metab 2006;16:233–244.

33 Holm L, Olesen JL, Matsumoto K, Doi T, Mizuno M, Alsted TJ, Mackey AL, Schwarz P, Kjaer M: Protein-containing nutrient supplementation following strength training enhances the effect on muscle mass, strength, and bone formation in postmenopausal women. J Appl Physiol 2008;105:274–281.

34 White KM, Bauer SJ, Hartz KK, Baldridge M: Changes in body composition with yogurt consumption during resistance training in women. Int J Sport Nutr Exerc Metab 2009;19:18–33.

35 Brown EC, DiSilvestro RA, Babakinia A, Devor ST: Soy versus whey protein bars: effects on exercise training impact on lean body mass and antioxidant status. Nutr J 2004;3:22.

36 Moore DR, Robinson MJ, Fry JL, Tang JE, Glover EI, Wilkinson SB, Prior T, Tarnopolsky MA, Phillips SM: Ingested protein dose response of muscle and albumin protein synthesis after resistance exercise in young men. Am J Clin Nutr 2009;89:161–168.

37 Weaver CM: Role of dairy beverages in the diet. Physiol Behav 2010;100:63–66.

38 Thomas DT, Wideman L, Lovelady CA: Effects of calcium and resistance exercise on body composition in overweight premenopausal women. J Am Coll Nutr 2010;29:604–611.

39 Josse AR, Atkinson SA, Tarnopolsky MA, Phillips SM: Diets higher in dairy foods and dietary protein support bone health during diet- and exercise-induced weight loss in overweight and obese premenopausal women. J Clin Endocrinol Metab 2012; 97:251–260.

40 Wagner G, Kindrick S, Hertzler S, DiSilvestro RA: Effects of various forms of calcium on body weight and bone turnover markers in women participating in a weight loss program. J Am Coll Nutr 2007;26: 456–461.

41 Major GC, Alarie FP, Dore J, Tremblay A: Calcium plus vitamin D supplementation and fat mass loss in female very low-calcium consumers: potential link with a calcium-specific appetite control. Br J Nutr 2009;101:659–663.

42 Layman DK, Evans E, Baum JI, Seyler J, Erickson DJ, Boileau RA: Dietary protein and exercise have additive effects on body composition during weight loss in adult women. J Nutr 2005;135:1903–1910.

43 Wallace BA, Cumming RG: Systematic review of randomized trials of the effect of exercise on bone mass in pre- and postmenopausal women. Calcif Tissue Int 2000;67:10–18.

44 Maestu J, Eliakim A, Jurimae J, Valter I, Jurimae T: Anabolic and catabolic hormones and energy balance of the male bodybuilders during the preparation for the competition. J Strength Cond Res 2010; 24:1074–1081.

Prof. Stuart M. Phillips
Department of Kinesiology, McMaster University
1280 Main Street West
Hamilton, ON L8S 4K1 (Canada)
Tel. +1 905 525 9140, ext. 24465, E-Mail phillis@mcmaster.ca

Lamprecht M (ed): Acute Topics in Sport Nutrition.
Med Sport Sci. Basel, Karger, 2013, vol 59, pp 104–112

Glycerol Use in Hyperhydration and Rehydration: Scientific Update

S.P. van Rosendal · J.S. Coombes

Human Performance Laboratory, School of Human Movement Studies, The University of Queensland, Brisbane, Qld., Australia

Abstract

Glycerol ingestion creates an osmotic drive that enhances fluid retention. The major practical applications for athletes are to either (i) hyperhydrate before exercise so that they have more fluid to be lost as sweat during subsequent performance, thereby delaying the progression of dehydration from becoming physiologically significant, or (ii) improve both the rate of rehydration and total fluid retention following exercise. Recently we showed that rehydration may be improved further by combining glycerol with intravenous fluids. Improvements in endurance time, time trial performance and total power and work output have been seen during exercise following glycerol-induced hyperhydration or rehydration. Another recent trial showed that the increased body weight associated with the extra fluid does not inadvertently affect running economy. Concerns that the haemodilution associated with the fluid retention in the vascular space may be sufficient to mask illegal doping practices by athletes led the World Anti-Doping Agency (WADA) to add glycerol to its list of prohibited substances in 2010. Recent evidence suggests that doses of $>0.032 \pm 0.010$ g/kg lean body mass (much lower than those required for rehydration) will result in urinary excretion that may be detectable, so athletes under the WADA jurisdiction should be cautious to limit their inadvertent glycerol intake.

Copyright © 2012 S. Karger AG, Basel

Glycerol was first investigated as a substrate to enhance fluid retention by Riedesel et al. [1] in 1987. Since then, 26 published studies have investigated the use of glycerol as a hyperhydrating agent, while another five have used glycerol to enhance rehydration (two with fluids given during exercise and three with fluids given post-exercise) and a further one study incorporated glycerol with a typical pre-race fluid regime rather than with a larger hyperhydrating fluid bolus [2]. The current scientific update will initially focus on the implications of the 2010 ruling by the World Anti-Doping Agency (WADA) and recent studies investigating glycerol use, before closing with a brief review of recommendations for glycerol use. A more comprehensive analysis of the hydration, cardiovascular, thermoregulatory and performance benefits of glycerol and the justifications supporting our recommendations can be found in our previous reviews [2, 3].

Ruling by the World Anti-Doping Agency

The most significant recent advancement regarding the use of glycerol by athletes has been the addition of glycerol to the WADA 2010 prohibited list under section S5 'Diuretics and Other Masking Agents' due to its effects as a plasma expander, making it prohibited at all times (in- and out-of-competition) [4]. Plasma expanders are prohibited under the WADA code due to their promoting fluid retention within the vascular space, thereby causing a haemodilution effect. Haemodilution may lower the concentration of actual doping substances or chemicals in the blood to a level that may be insufficient to be detected, or it may aid an athlete trying to avoid detection for doping practices by surreptitiously masking states of artificially increased haemoglobin/haematocrit concentration (e.g. from erythropoietin use or autologous blood transfusions), to a level that could be considered natural. A recent study examined the haemodilution effects of a combined creatine/glycerol protocol that induced a total body water increase of 0.67 ± 0.17 l [5]. They found a non-significant ($p = 0.07$) reduction in haemoglobin concentration of approximately 0.4 g/l and a significant ($p < 0.05$) reduction in haematocrit of $1.36 \pm 0.73\%$ following hyperhydration [5]. Therefore, it is plausible that hyperhydration could be used to avoid detection following doping.

One concern with the prohibition of glycerol is that it occurs both naturally in the body as the backbone of the triglyceride molecule, and in many foods and other products that athletes will consume inadvertently. The concentration of serum glycerol at rest is approximately 0.05–0.1 mmol/l, however it can be elevated to levels approaching 0.3–0.4 mmol/l during periods of increased lipolysis, for example in response to prolonged exercise or caloric restriction [6]. The ruling by WADA was clarified in the Explanatory Notes on the 2011 Prohibited List (released 18 September 2010): '*The prohibition of glycerol is not intended to prevent the ingestion of this substance which is commonly found in a variety of foodstuffs and toiletries. Such use will not cause a competitor to test positive for this Prohibited Substance*' [7]. In response to this ruling by WADA, a recent study by Nelson et al. [8] aimed to investigate the plasma glycerol concentration that coincides with urinary glycerol excretion. Subjects consumed five different glycerol doses (0.025, 0.05, 0.10, 0.15, or 0.20 g/kg of lean body mass) made as a 5% glycerol solution. Importantly, these doses are all much lower than the dose required to induce hyperhydration (generally regarded as ≥ 1.0 g/kg body weight [1]). The study monitored plasma and urinary glycerol concentrations to determine both the oral dose and the specific plasma glycerol concentration that coincided with an increase in urine glycerol concentration. They conducted a regression analysis which showed that a glycerol dose of $>0.032 \pm 0.010$ g/kg lean body mass and plasma glycerol concentrations $>0.327 \pm 0.190$ mmol/l would be associated with urinary glycerol excretion [8], a figure much lower than the 1.1–1.6 mmol/l reported from animal studies (e.g. Tourtellotte et al. [9]). They therefore concluded that it is possible to detect prohibited use of glycerol as a masking agent through urinary analysis [8].

Although no specific plasma or urinary level has been released by WADA as a level that constitutes a cut-off for doping, it is recommended that glycerol should not be used in the doses required to induce hyperhydration by any competitive athlete whose sport comes under the jurisdiction of the WADA code. Importantly, it is worth recalling at this juncture that a plasma level approaching 0.3–0.4 mmol/l may naturally occur during periods of increased lipolysis [6]. Thus, in athletes exercising for prolonged periods (e.g. in ultra-endurance events), it is conceivable that only minimal exogenous glycerol intake may be required to increase plasma levels sufficiently to detect an increase in urinary excretion and additional caution may be required by these athletes with regard to incidental exogenous glycerol intake to avoid detection in their urine.

Potential Negative Effects of Glycerol Hyperhydration

In a recent review article, we raised the possibility that the body weight gain associated with glycerol hyperhydration may actually have a detrimental effect on performance during weight-bearing activity of a relatively short duration or in non-stressful environmental conditions [3]. We considered this in the context that the force required to support body weight is the primary factor determining the metabolic cost of walking or running. Grabowski et al. [10] showed that 28 and 45% of the metabolic cost of walking went to supporting body weight and accelerating the centre of gravity, respectively. This contribution rises even further when running, during which 74% of the net metabolic cost of the activity is required to support body weight [11]. We noted that some results possibly support this assertion, with one glycerol hyperhydration study showing that 10-m sprint times tended to be higher immediately following hyperhydration when subjects were ~1 kg heavier, [12] while another showed a non-significant (5%) reduction in exercise time during a short incremental field test when athletes were ~0.6 kg heavier [13].

The possibility that glycerol hyperhydration could detrimentally effect performance was recently studied in relation to running economy. Running economy is typically defined as the oxygen consumption (Vo_2) or energy demand for a given submaximal running velocity [14]. The importance of running economy is highlighted by the fact that a strong association exists between running economy and distance (endurance) running performance [14]. In fact, running economy has been shown to be a better predictor of performance than maximal oxygen uptake ($Vo_{2\ max}$) in elite runners [14]. It has been hypothesised that reductions in body weight due to dehydration might lower the energy cost of movement with a relative increase in power output [14]. For example, if dehydration reduces the body weight and therefore the absolute oxygen cost of running by a specific level (e.g. 4%), but the absolute power output by less than this (e.g. 3%), then power output per kilogram could conceivably be increased (so long as the individual can tolerate the hyperthermia associated with the activity) [15]. This theory is supported somewhat by the finding that athletes who finish highest in endurance events (e.g. marathon running) also tend to lose the most body weight during the race [16]. It has been proposed that this may occur because the reduced body weight (i.e. fluid loss) lowers the oxygen cost of movement [15]. One may therefore hypothesise that the increase in body weight associated with glycerol ingestion could lead to reduced running efficiency and an increased Vo_2 for the same submaximal workload.

The influence of glycerol hyperhydration on running economy was recently investigated by Beis et al. [17]. They hyperhydrated 15 trained male distance runners with a protocol combining a week long supplementation of creatine and glycerol as used previously by Easton et al. [18]. Body mass and total body water increased by 0.9 ± 0.4 kg and 0.71 ± 0.42 l following hyperhydration, respectively [17]. Subjects then ran on a treadmill at 60% $Vo_{2\,max}$ in both cool (10°C, 70% relative humidity) and hot and humid (35°C, 70% relative humidity) environmental conditions [17]. They found that glycerol hyperhydration was associated with improved thermoregulatory responses to exercise in the heat (heart rate and core temperature were significantly lower during exercise in the hot/humid environment following supplementation, compared to pre-supplementation). However, no significant differences in any of the respiratory variables (Vo_2, Vco_2, respiratory exchange ratio) were found between the pre- and post-supplementation exercise trials, indicating that the increase in body weight did not negatively impact running economy [17]. These findings support the use of hyperhydration in preparation for endurance running, especially when exercise is undertaken in hot and humid conditions [17]. It is worth noting here that the exercise intensity used in this study (60% $Vo_{2\,max}$) is lower than most competitive events and results showing similar findings at faster running speeds more typical of sporting competition (e.g. >85% $Vo_{2\,max}$) would be beneficial [17].

Latest Advancements in Glycerol Use during Rehydration

In a recent study we investigated whether oral glycerol ingestion would enhance rehydration when combined with intravenous (IV) fluid following exercise-induced dehydration [19]. Endurance trained men were dehydrated to –4% body weight before being rehydrated via four different fluid regimes (in a randomised cross-over design). The fluid regimes were (1) *oral* (combination of oral sports drink and water) without glycerol; (2) *oral glycerol* (combination of oral sports drink and water with oral glycerol (1.5 g/kg body weight)); (3) *IV* (half IV fluid (0.9% NaCl), half oral fluid (combination of sports drink and water) without glycerol), and (4) *IV with oral glycerol* (half IV fluid (0.9% NaCl), half oral fluid (combination of sports drink and water) with oral glycerol (1.5 g/kg body weight)). Subjects then completed a performance test (30 min of cycling at 85% ventilatory threshold followed by a 40-km cycling time trial) in the heat (35°C, 70% relative humidity). The major findings were: (1) IV fluid had the greatest effect on plasma volume, however the addition of glycerol to the rehydration beverages increased the restoration of plasma volume significantly more than the corresponding fluid regime without glycerol, so that plasma volume was highest in *IV with oral glycerol > IV > oral glycerol > oral;* (p < 0.01); (2) compared to *oral* rehydration, subsequent 40-km time trial performance was improved following the *IV* (3.5%), *oral glycerol* (3.7%) and *IV with oral glycerol* (4.1%) rehydration regimes; (3) *IV with oral glycerol* resulted in the greatest net fluid balance following rehydration and exercise performance, and (4) there were no differences in cardiovascular (heart rate), thermoregulatory (tympanic/skin temperatures, sweat rate), metabolic (blood lactate), or subjective (rating of perceived exertion, thermal stress) measures between rehydration conditions [19]. In addition, we found that glycerol ingestion had no influence on levels of antidiuretic hormone, aldosterone or cortisol levels during rehydration or subsequent performance (unpubl. data). Importantly, IV rehydration has also recently been banned by WADA, so while the results from this latest trial indicates that combining glycerol with IV fluids can further enhance rehydration, athletes should again be discouraged from undertaking the practice if they fall under the jurisdiction of WADA.

Considerations for Glycerol Use in Hyperhydration

While off limits to elite athletes who fall under the WADA code and who must therefore follow its list of banned substances, glycerol remains an attractive option for use by sub-elite endurance athletes who are not required to submit to drug testing. It remains attractive because it has been shown to provide hydration and performance benefits across many studies, while being a very safe substrate when taken at the doses used in hydration research. The following factors should be considered by anyone intending on using glycerol and have been described in detail elsewhere [2]. Therefore only a brief description of each is provided here.

Dose
Robergs and Griffin [6] found that a glycerol dose of 1.0–1.5 g/kg body weight will be required to elevate plasma glycerol levels to 15 mmol/l, the concentration above which plasma glycerol levels stabilise. Therefore, doses higher than this provide no additional osmotic gradient for fluid retention. Practical results have supported these pharmacokinetic findings. The average dose of glycerol used across all hyperhydration studies is 1.1 g/kg body weight [20], while doses of 1.2 g/kg body weight have been associated with the maximal increases to total body water [2].

Fluid Volume
In a recent meta-analysis, Goulet et al. [20] found that a volume of 26 ml/kg body weight would provide maximal fluid retention. Such a volume (e.g. 1,820 ml for a

70-kg athlete) may be difficult for some individuals to tolerate. Siegler et al. [13] used a regime whereby athletes consumed 500 ml of a glycerol solution within a 30-min period before a 60-min training session, plus another 500 ml halfway through training. This was sufficient to attenuate the body weight and plasma volume losses during exercise [13]. Therefore, if an athlete is unable to tolerate the large volumes of fluid and glycerol that are recommended to induce hyperhydration, then they may wish to try smaller volumes of fluid with glycerol shortly before starting exercise.

Type of Fluid
Most glycerol hyperhydration studies have used aspartame-sweetened water (to mask the taste of glycerol in double-blind trials) as the hyperhydrating beverage and have consistently induced hyperhydration. Given the rationale that sports drinks may provide a better hydration potential than water alone, while also providing sodium to attenuate hyponatraemia, one could hypothesise that consuming carbohydrate-electrolyte beverages with glycerol will further enhance the hyperhydration benefits of glycerol. No study has yet compared the inclusion of glycerol in plain water compared to a sports drink.

Timing of Fluid
Pre-exercise fluid and glycerol administration is achieved by one of two regimes. Athletes either rapidly ingest a small concentrated bolus of glycerol solution followed by the remaining fluid bolus over a longer duration (e.g. 120–150 min), or they mix the glycerol dose in the entire hyperhydration fluid bolus to spread the glycerol consumption over a longer period of time [2]. Fluid retention tends to be greatest when the glycerol and fluid is ingested over 60–150 min [2].

Duration of Fluid Retention
Glycerol maintains hyperhydration for periods of 4 h or more. After 2 and 3 h, fluid retention of 60–80 and 45–60% respectively have been reported (compared with 40–60 and 10–30%, respectively, with water-induced hyperhydration) [2].

Glycerol Use in Rehydration

The efficacy of glycerol use in rehydration depends on factors related to the level of dehydration achieved during exercise (and therefore whether the athlete hyperhydrated before exercise, the environmental conditions, the exercise duration and intensity and the athlete's sweat rate) and the time available for rehydration before another exercise bout. To date, only three studies have explored the role of glycerol in rehydration. Each of these found that beverages containing glycerol were associated with significantly more rapid and complete restoration of plasma volume than water alone. However, if dehydration during exercise is not severe enough to detrimentally affect performance, then the addition of glycerol to rehydration beverages is unlikely to convey benefit. Similarly, if sufficient time is available between exercise bouts to replenish the fluid compartments with normal oral intake, then the inclusion of glycerol is again unlikely to convey a benefit. However, if an athlete is performing several intense, long duration bouts of exercise, with a short period in between for rehydration, then glycerol consumption may be of benefit to enhance rehydration. A dose of 1.0 g/kg body weight in each 1.5 l of fluid consumed will provide a similar regime to those used for hyperhydration [2].

Performance Benefits

Exercise performance following glycerol hyperhydration or rehydration has been evaluated in 18 studies, of which 11 showed significant performance improvements in the glycerol trials [3]. Performance improvements include increased endurance time to exhaustion (by up to 24%), improved time trial performance (by up to 4.1%) and increased power or work (by up to 5%) [3]. While other studies have found no performance benefits, none have shown a significant performance decrement following glycerol hyperhydration or rehydration.

Thermoregulatory and Cardiovascular Benefits

In studies that have shown benefits, the improvements have been associated with cardiovascular, thermoregulatory and subjective benefits such as increased plasma volume and sweat rate, and reduced core temperature and ratings of perceived exertion [3].

Side Effects

The incidence of side effects associated with glycerol consumption at doses used to enhance hydration is very low. Some studies have reported that a small number of subjects complained of mild symptoms such as nausea, gastrointestinal discomfort/ bloatedness, dizziness, light-headedness or headaches within 24 h of consuming the glycerol. The risk of more serious side effects such as cerebral dehydration would potentially increase if glycerol accumulates in the body as a result of multiple large glycerol doses over extended periods of time, especially when dehydrated. Thus, hyperhydrating with high doses of glycerol for extended periods (i.e. >4 h) is not recommended. There are also certain populations for whom glycerol ingestion should be avoided, including individuals with diabetes, renal disease, migraine and headache disorders, cardiovascular disease, liver disorders, and in pregnancy.

Conclusion

Recent studies investigating glycerol use in hyperhydration and rehydration have focussed on (1) the levels of glycerol required to result in urinary excretion, (2) the impact of hyperhydration on running economy, and (3) the combination of IV fluids and oral glycerol as a method to maximise rehydration. Findings from these studies indicate that (1) urinary excretion of glycerol occurs with doses that are only marginally higher than levels that can occur naturally during periods of prolonged lipolysis and well below those required to induce hyperhydration, (2) submaximal running economy is not adversely affected by the gain in body weight associated with hyperhydration, and (3) combining IV fluids with oral glycerol may maximise rehydration.

Disclosure Statement

The authors have no conflicts of interest or sources of funding to disclose.

References

1 Riedesel ML, Allen DY, Peake GT, Al-Qattan K: Hyperhydration with glycerol solutions. J Appl Physiol 1987;63:2262–2268.

2 van Rosendal SP, Osborne MA, Fassett RG, Coombes JS: Guidelines for glycerol use in hyperhydration and rehydration associated with exercise. Sports Med 2010;40:113–129.

3 van Rosendal SP, Osborne MA, Fassett RG, Coombes JS: Physiological and performance effects of glycerol hyperhydration and rehydration. Nutr Rev 2009;67:690–705.

4 WADA: 2012 List of Prohibited Substances and Methods, 2012 (cited 2012 January 30); available from http://list.wada-ama.org/

5 Mohammad YA: Detection of blood doping in athletes (masking substances and methods); thesis, Glasgow 2010.

6 Robergs RA, Griffin SE: Glycerol. Biochemistry, pharmacokinetics and clinical and practical applications. Sports Med 1998;26:145–167.

7 WADA: Explanatory Notes on the 2011 Prohibited List, 2010 (cited 2012 January 30); available from http://www.wada-ama.org/Documents/World_Anti-Doping_Program/WADP-Prohibited-list/To_be_effective/WADA_Explanatory_Notes_Prohibited_List_EN.pdf

8 Nelson JL, Harmon ME, Robergs RA: Identifying plasma glycerol concentration associated with urinary glycerol excretion in trained humans. J Anal Toxicol 2011;35:617–623.

9 Tourtellotte WW, Reinglass JL, Newkirk TA: Cerebral dehydration action of glycerol. I. Historical aspects with emphasis on the toxicity and intravenous administration. Clin Pharmacol Ther 1972;13:159–171.

10 Grabowski A, Farley CT, Kram R: Independent metabolic costs of supporting body weight and accelerating body mass during walking. J Appl Physiol 2005;98:579–583.

11 Teunissen LP, Grabowski A, Kram R: Effects of independently altering body weight and body mass on the metabolic cost of running. J Exp Biol 2007;210:4418–4427.

12 Magal M, Webster MJ, Sistrunk LE, Whitehead MT, Evans RK, Boyd JC: Comparison of glycerol and water hydration regimens on tennis-related performance. Med Sci Sports Exerc 2003;35:150–156.

13 Siegler JC, Mermier CM, Amorim FT, Lovell RJ, McNaughton LR, Robergs RA: Hydration, thermoregulation, and performance effects of two sport drinks during soccer training sessions. J Strength Cond Res 2008;22:1394–1401.

14 Saunders PU, Pyne DB, Telford RD, Hawley JA: Factors affecting running economy in trained distance runners. Sports Med 2004;34:465–485.

15 Coyle EF: Fluid and fuel intake during exercise. J Sports Sci 2004;22:39–55.

16 Zouhal H, Groussard C, Minter G, Vincent S, Cretual A, Gratas-Delamarche A, et al: Inverse relationship between percentage body weight change and finishing time in 643 forty-two-kilometre marathon runners. Br J Sports Med 2011;45:1101–1105.

17 Beis LY, Polyviou T, Malkova D, Pitsiladis YP: The effects of creatine and glycerol hyperhydration on running economy in well-trained endurance runners. J Int Soc Sports Nutr 2011;8:24.

18 Easton C, Turner S, Pitsiladis YP: Creatine and glycerol hyperhydration in trained subjects before exercise in the heat. Int J Sport Nutr Exerc Metab 2007; 17:70–91.

19 van Rosendal SP, Strobel NA, Osborne MA, Fassett RG, Coombes JS: Performance benefits of rehydration with intravenous fluid and oral glycerol. Med Sci Sports Exerc 2012;44:1780–1790.

20 Goulet ED, Aubertin-Leheudre M, Plante GE, Dionne IJ: A meta-analysis of the effects of glycerol-induced hyperhydration on fluid retention and endurance performance. Int J Sport Nutr Exerc Metab 2007;17:391–410.

Prof. Jeff S. Coombes
School of Human Movement Studies
The University of Queensland
St Lucia, QLD 4072 (Australia)
Tel. +61 7 33656767, E-Mail jcoombes@uq.edu.au

Lamprecht M (ed): Acute Topics in Sport Nutrition.
Med Sport Sci. Basel, Karger, 2013, vol 59, pp 113–119

Salt and Fluid Loading: Effects on Blood Volume and Exercise Performance

Ricardo Mora-Rodriguez · Nassim Hamouti

University of Castilla-La Mancha, Exercise Physiology Laboratory, Toledo, Spain

Abstract

During prolonged exercise, fluid and salt losses through sweating reduce plasma volume which leads to heart rate drift in association with hyperthermia and reductions in performance. Oral rehydration with water reduces the loss of plasma volume and lessens heart rate drift and hyperthermia. Moreover, the inclusion of sodium in the rehydration solution to levels that double those in sweat (i.e., around 90 mmol/l Na^+) restores plasma volume when ingested during exercise, and expands plasma volume if ingested pre-exercise. Pre-exercise salt and fluid ingestion with the intention of expanding plasma volume has received an increasing amount of attention in the literature in recent years. In four studies, pre-exercise salt and fluid ingestion improved performance, measured as time to exhaustion, either during exercise in a thermoneutral or in a hot environment. While in a hot environment, the performance improvements were linked to lowering of core temperatures and heart rate, the reasons for the improved performance in a thermoneutral environment remain unclear. However, when ingesting pre-exercise saline solutions above 0.9% (i.e., >164 mmol/l Na^+), osmolality and plasma sodium increase and core temperature remain at dehydration levels. Thus, too much salt counteracts the beneficial effects of plasma volume expansion on heat dissipation and hence in performance. In summary, the available literature suggests that pre-exercise saline ingestion with concentrations not over 164 mmol/l Na^+ is an ergogenic aid for subsequent prolonged exercise in a warm or thermoneutral environment.

While the physiological benfits of water replenishment during and after prolonged exercise are well recognized, the physiological role of salt replacement is debatable. Sodium in blood exerts osmotic forces that defend plasma volume during prolonged exercise inducing dehydration. One of the prominent adaptations to chronic exercise in a hot climate (i.e., acclimation) is to reduce sodium excretion in sweat [1]. By doing so, more sodium is kept in the blood, increasing osmotic forces that help to maintain blood volume during progressive dehydration. The role of sodium in blood volume maintenance is masterly illustrated in a recent experiment comparing cystic fibrosis patients who excrete a lot of sodium in sweat (133 mmol/l) with control subjects with average sweat sodium of 44 mmol/l [2]. Both groups exercised in a hot environment while dehydrating up to 3%. Due to their large sweat Na^+ losses, the patients finished the exercise with a lower blood sodium concentration than the controls (146 vs. 150 mmol/l, respectively). With less sodium in blood to exert osmotic forces, the cystic fibrosis group had larger plasma volume reductions during exercise than the control subjects.

While curtailing sodium losses could benefit the cardiovascular system by reducing blood volume losses, adding sodium to the blood holds the promise to maintain blood volume during prolonged exercise. This is the main focus of this review. We will present current information about the acute use of water and salt as a nutritional aid that can help cardiovascular function. The impact of salt and water loading on reducing cardiovascular and thermal strain and its consequences on exercise performance will also be discussed. Most of the review will deal with oral ingestion although intravenous delivery of saline solutions is also presented with the aim of clarifying the physiological mechanisms. Human sweat during exercise contains ~45 mmol/l of Na^+. Thus, studies using drinks with sodium concentrations below that range of concentrations (i.e., sports drinks) are not considered salt loading and are beyond the scope of this review. Lastly, we focused on studies in a normotensive population with normal renal function and appropriate ADH secretion. Salt and water loading is obviously discouraged in hypertensive populations.

Saline Delivery after Dehydration but prior to a Subsequent Exercise Bout

In several studies saline solutions have been administered to manipulate (i.e., restore of expand) plasma volume. Fortney et al. [3] rehydrated with saline subjects that had lost 200 ml of their plasma volume after exercise in the heat combined with water restriction. They used intravenous (IV) infusion of 3% saline (0.4 ml/kg body mass/min) to successfully restore plasma volume. However, due to the high osmolality of the infusate (i.e., 1,026 mosm/kg H_2O), plasma osmolality remained at dehydrated values and thermoregulation did not benefit from intravascular rehydration. A series of studies have followed (reviewed by van Rosendal et al. [4]), using isotonic (0.9%; ~308 mosm/kg H_2O) or hypotonic (0.45%; ~154 mosm/kg H_2O) IV saline infusion to rehydrate subjects that have lost fluid during prolonged exercise in the heat. In general, these IV saline solutions restored plasma volume to euhydrated conditions without differences between the iso- and hypotonic saline [5]. The acute expansion of blood volume with the IV saline infusion seemed to restore central venous pressure improving heat dissipation. In fact, in these studies, core temperature was lower during subsequent exercise than when subjects did not receive saline infusions.

The main goal of the above-cited studies was to compare the rehydration effects of IV vs. oral solutions and thus 0.45% saline solutions were also ingested. The main conclusion of these studies is, that although IV rehydration seemed to be faster at restoring plasma volume, the cardiovascular, thermoregulatory and performance benefits during subsequent exercise were similar to when rehydrating orally [4]. While the mode of delivery of the saline (IV vs. oral) was nicely addressed in these studies, no conclusions can be derived about the effects of adding salt to a rehydration fluid since a water ingestion control trial was not included.

Saline Delivery during Dehydrating Exercise

We found three articles investigating the role of IV saline infusion delivered during a dehydrating exercise on the cardiovascular, thermoregulatory and performance responses to exercise. In two of them, saline was infused while pedaling in a hot environment (30°C) while the other study was held in a thermoneutral environment but at higher workload (i.e., 84% vs. 60–65% $Vo_{2\,max}$) and thus all of them caused moderate levels of heat accumulation. The three studies coincided in that infusion of isotonic saline (0.9%; 0.3–0.9 ml/kg body weight/min) during exercise restored plasma volume to pre-exercise levels [6–8]. The infusions suppressed the gradual drift in heart rate and reduced the hyperthermia observed when no fluid was delivered. IV infusion of isotonic saline neither raised plasma sodium [6] nor blood osmolality [7]. Interestingly, performance measured by endurance time was not improved by the IV saline infusion [6]. However, there was a wide spread of times to fatigue (8–42 min) among the subjects participating in that study which together with the low reliability of time to exhaustion to measure performance [20] makes those results inconclusive.

We found three studies using saline ingestion during dehydrating exercise. In one study [9], subjects ingested slightly hypertonic saline (1%; ~342 mosm/kg H_2O) as a rehydration fluid prior to and during exercise avoiding dehydration (i.e., <0.5% body weight loss). The ingestion of saline raised blood osmolality to the levels of when no fluid was ingested (i.e., dehydration trial). However, saline ingestion maintained plasma volume at pre-exercise levels, well above the dehydration trial and somewhat above the water ingestion trial. Despite this positive cardiovascular effect of saline ingestion, aural temperature increased above the water trial to levels similar to the dehydration trial. Similar to what happened with hypertonic IV saline infusions [3], the ingestion of hypertonic saline negated the thermoregulatory benefits that may bring about the expansion of plasma volume.

Finally, in two experiments by Sanders et al. [10, 11], subjects ingested a 400-ml saline bolus before exercise and 100- and 150-ml aliquots every 10 min during 3–4 h exercise at 55–65% of $Vo_{2\,max}$. In one study held in a 32°C environment [10], the fluid ingested only replaced 50% of the fluid losses and there was no difference between ingesting water or a saline solution of 100 mmol Na^+ (~0.58% saline). For instance, plasma volume and even plasma sodium concentration were equally maintained with water than with saline ingestion and heart rate drift was similarly attenuated. The low sodium dose and volume ingested was possibly not enough to act as a plasma volume expander. In their second experiment, fluid intake matched sweat losses during prolonged exercise in a themoneutral environment (i.e., 20°C). On this occasion, the ingestion of a 100 mmol/l Na^+ solution maintained plasma volume (i.e., extracellular fluid) better than when ingesting a 5-mmol/l Na^+ solution. This was achieved by increased fluid retention with reduced urine production that however did not affect the heart rate drift or increase in rectal temperature. Due to the low exercise intensity and environmental heat stress, heart rate drifted only 6% during 4 h of pedaling and rectal temperature 1.8°C from the 15-min value [11]. It is possible that in a hotter environment the effects of saline ingestion could have been more evident.

Saline Delivery Pre-Exercise: Effects on Plasma Volume

Greenleaf et al. [12] investigated the optimal composition of sodium beverages for increasing plasma volume prior to exercise in euhydrated individuals. Using a combination of sodium chloride and sodium citrate, they delivered drinks containing 55 mmol/l of Na^+, however hypertonic (i.e., 365 mosm/kg H_2O) and 164 mmol/l of Na^+ but hypotonic (i.e., 253 mosm/kg H_2O). After a drinking and resting period (i.e., ~100 min), they found that the higher the sodium concentration, the higher the increase in plasma volume (i.e., 5 vs. 8%, respectively). They concluded that the sodium concentration of the solution appears to be more important than its osmolality to expand plasma volume. These results were confirmed by the same research group in another study using a similar drink formulation [13].

Since the work of Greenleaf and co-workers, other investigators have used oral saline solutions of 164 mmol/l of Na^+ to induce pre-exercise hypervolemia and to study its effects in performance during subsequent exercise under different environmental conditions [14–17]. Unlike Greenleaf et al. [12], these authors found lower levels of plasma volume expansion (i.e., 3–4.5% above resting values) despite ingesting a similar Na^+ solution of 164 mmol/l. We have recently found a similar low level of plasma volume expansion after ingestion of 10 ml/kg body mass of 164 mmol/l Na^+ solution (1% expansion; Hamouti et al., unpubl. data). This disparity in the level of plasma volume expansion between Greenleaf et al. and other authors may be due to the lower aerobic fitness level among their participants ($Vo_{2 max}$ of 40 vs. 55 ml O_2/kg/min in the other studies). Aerobically fit subjects are hypervolemic as a consequence of endurance training adaptations [18]. It is then possible that training reduces the amount of plasma volume available to be expanded since they are close to the ceiling for expansion. Figure 1 depicts what level of plasma expansion could be expected when ingesting saline solutions with regards to the aerobic fitness level of the subject.

Saline Delivery Pre-Exercise: Effects on Performance

We could find only two studies in a thermoneutral environment reporting effects of pre-exercise saline ingestion on exercise performance. Greenleaf et al. [13] found

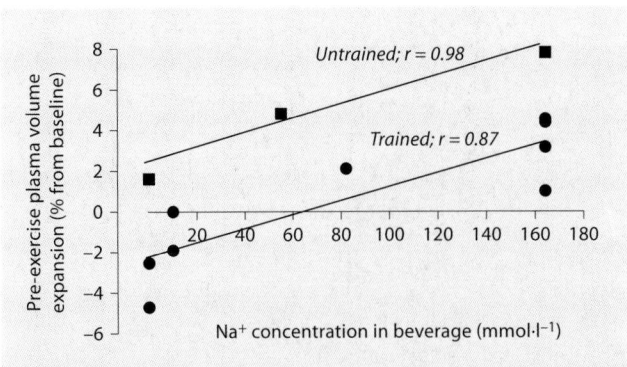

Fig. 1. Relationship between sodium concentration in the beverages from several studies and the percent of pre-exercise plasma volume expansion. Data is separated in untrained ($Vo_{2\,max}$ 40 ml O_2/kg/min) and trained individuals ($Vo_{2\,max}$ 55 ml O_2/kg/min). Each point represents one experimental trial [12–17].

that, compared with a moderate sodium solution (i.e., 55 mmol/l of Na^+), pre-exercise ingestion of 164 mmol/l Na^+ solution improved cycling time to exhaustion by 24%. However, this improvement was not associated with changes in the cardiovascular (i.e., heart rate) or thermoregulatory responses (i.e., core temperature or total whole body sweat rate) regardless of higher blood availability. Coles and Luetkemeier [14] found an 8% improvement in time-trial performance upon a 3% pre-exercise blood volume expansion with a 164-mmol/l Na^+ drink compared to a non-sodium placebo drink. Similar to Greenleaf et al. [12, 13], they did not find differences in heart rate, core temperature, rate of perceived exertion or total body sweat rate. It is then intriguing what could have caused the improvements in performance. It is possible that the increased plasma volume may have permitted an increase in $Vo_{2\,max}$ and thus in time to exhaustion [19]. Supporting this possibility, Coles and Luetkemeier [14] reported that the individuals with the lower maximal aerobic fitness level (and probably lower initial blood volume) had the greatest increase in performance upon plasma volume expansion with saline ingestion.

In a hot environment, only two studies report performance results after pre-exercise saline ingestion [15, 16]. There is one other study on the heat using saline ingestion to expand plasma volume, but performance is not reported [17]. In agreement with the thermoneutral studies just discussed, performance time to exhaustion was improved by 22% after the ingestion of a 164 mmol/l Na^+ in comparison to the ingestion of a 10 mmol/l Na^+ solution. These rather large improvements in performance occurred despite moderate pre-exercise plasma volume expansion (i.e., 4.5% [15, 16]). In the case of exercise in the heat, the performance effects could be related to the lower core temperature and perceived exertion associated with the increased plasma volume. Using pre-exercise ingestion of a 82 mmol/l Na^+ solution, we have

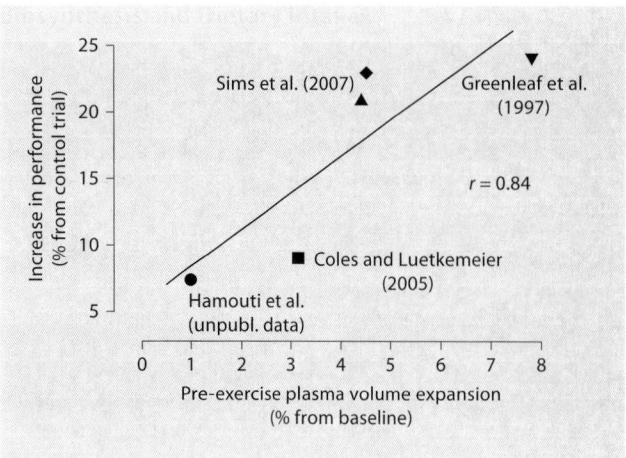

Fig. 2. Relationship between the percent of pre-exercise plasma volume expansion induced by the ingestion of 164 mmol/l Na⁺ solution (10 ml/kg body mass) and the percent of increase in performance with respect to the control trial being the ingestion of the same volume of a low sodium solution or plain water. Each point represents one study.

recently found similar findings of improved performance but using a cycling time trial instead of the less reliable time to exhaustion performance test [Hamouti et al., unpubl. data].

Conclusion

There is conflicting information regarding the performance effects of pre-exercise saline ingestion. Factors like the sodium concentration of the solution and the level of pre-exercise plasma volume expansion seems to play an important role in endurance performance. In figure 2, we show available data which suggest that the higher the pre-exercise plasma volume expansion, the higher the increase in endurance performance. Again, a large plasma volume expansion is only possible in subjects with a low aerobic fitness level and thus the ergogenic power of pre-exercise saline ingestion for a trained athlete may be moderate in a thermoneutral environment and somewhat higher in a hot environment.

Disclosure Statement

The authors have no conflicts of interest to disclose.

References

1   Buono MJ, Ball KD, Kolkhorst FW: Sodium ion concentration vs. sweat rate relationship in humans. J Appl Physiol 2007;103:990–994.

2   Brown MB, McCarty NA, Millard-Stafford M: High-sweat Na⁺ in cystic fibrosis and healthy individuals does not diminish thirst during exercise in the heat. Am J Physiol Regul Integr Comp Physiol 2011;301:R1177–R1185.

3   Fortney SM, Wenger CB, Bove JR, Nadel ER: Effect of hyperosmolality on control of blood flow and sweating. J Appl Physiol 1984;57:1688–1695.

4   Van Rosendal SP, Osborne MA, Fassett RG, Lancashire B, Coombes JS: Intravenous versus oral rehydration in athletes. Sports Med 2010;40:327–346.

5   Kenefick RW, Maresh CM, Armstrong LE, Riebe D, Echegaray ME, Castellani JW: Rehydration with fluid of varying tonicities: effects on fluid regulatory hormones and exercise performance in the heat. J Appl Physiol 2007;102:1899–1905.

6   Deschamps A, Levy RD, Cosio MG, Marliss EB, Magder S: Effect of saline infusion on body temperature and endurance during heavy exercise. J Appl Physiol 1989;66:2799–2804.

7   Fortney SM, Vroman NB, Beckett WS, Permutt S, LaFrance ND: Effect of exercise hemoconcentration and hyperosmolality on exercise responses. J Appl Physiol 1988;65:519–524.

8   Nose H, Mack GW, Shi XR, Morimoto K, Nadel ER: Effect of saline infusion during exercise on thermal and circulatory regulations. J Appl Physiol 1990;69: 609–616.

9   Harrison MH, Edwards RJ, Fennessy PA: Intravascular volume and tonicity as factors in the regulation of body temperature. J Appl Physiol 1978; 44:69–75.

10 Sanders B, Noakes TD, Dennis SC: Water and electrolyte shifts with partial fluid replacement during exercise. Eur J Appl Physiol Occup Physiol 1999;80: 318–323.

11 Sanders B, Noakes TD, Dennis SC: Sodium replacement and fluid shifts during prolonged exercise in humans. Eur J Appl Physiol 2001;84:419–425.

12 Greenleaf JE, Looft-Wilson R, Wisherd JL, et al: Hypervolemia in men from fluid ingestion at rest and during exercise. Aviat Space Environ Med 1998;69:374–386.

13 Greenleaf JE, Looft-Wilson R, Wisherd JL, McKenzie MA, Jensen CD, Whittam JH: Pre-exercise hypervolemia and cycle ergometer endurance in men. Biol Sport 1997;14:103–114.

14 Coles MG, Luetkemeier MJ: Sodium-facilitated hypervolemia, endurance performance, and thermoregulation. Int J Sports Med 2005;26:182–187.

15 Sims ST, Rehrer NJ, Bell ML, Cotter JD: Pre-exercise sodium loading aids fluid balance and endurance for women exercising in the heat. J Appl Physiol 2007;103:534–541.

16 Sims ST, van Vliet L, Cotter JD, Rehrer NJ: Sodium loading aids fluid balance and reduces physiological strain of trained men exercising in the heat. Med Sci Sports Exerc 2007;39:123–130.

17 Nelson MD, Stuart-Hill LA, Sleivert GG: Hypervolemia and blood alkalinity: effect on physiological strain in a warm environment. Int J Sports Physiol Perform 2008;3:501–515.

18 Green HJ, Sutton JR, Coates G, Ali M, Jones S: Response of red cell and plasma volume to prolonged training in humans. J Appl Physiol 1991;70: 1810–1815.

19 Coyle EF, Hopper MK, Coggan AR: Maximal oxygen uptake relative to plasma volume expansion. Int J Sports Med 1990;11:116–119.

20 Jeukendrup A, Saris WHM, Brouns F, et al: A new validated endurance performance test. Med Sci Sports Exerc 1996;28:266–270.

Ricardo Mora-Rodriguez
University of Castilla-La Mancha
Avda Carlos III, s/n, ES–45071 Toledo (Spain)
Tel. +34 925 26 88 00, ext. 5510
E-Mail Ricardo.Mora@uclm.es

Lamprecht M (ed): Acute Topics in Sport Nutrition.
Med Sport Sci. Basel, Karger, 2013, vol 59, pp 120–126

Milk Protein and the Restoration of Fluid Balance after Exercise

Lewis James

School of Sport, Exercise and Health Sciences, Loughborough University, Loughborough, UK

Abstract

Sweat is produced during exercise to help dissipate some of the extra heat produced due to an increase in metabolic rate. Inadequate drink ingestion during exercise means athletes finish exercise hypohydrated and when the time between exercise bouts is short, effective rehydration strategies will be necessary to prevent subsequent performance impairment. For complete rehydration, drink volume must be sufficient to replace sweat losses as well as the additional water losses during recovery. Once a sufficient volume of drink is ingested it is the drink composition that dictates the rehydration success of the drink. It is well known that addition of sodium and some other nutrients to rehydration drinks enhances fluid balance restoration after exercise, but the effects of milk proteins have been less well documented. Skimmed milk is an effective post-exercise rehydration solution and enhances the restoration of fluid balance after exercise-induced dehydration to a greater extent than a carbohydrate-electrolyte sports drink. Whilst there are a number of factors in skimmed milk that might be responsible for this enhancement of rehydration, it appears that some of the effect is due to the milk protein, as milk protein has been shown to be more effective for post-exercise rehydration than an isoenergetic amount of carbohydrate. Whilst the effects of whey protein on post-exercise rehydration are equivocal, whey protein addition to a carbohydrate-electrolyte rehydration solution certainly does not impair rehydration. Therefore, in situations where protein ingestion after exercise might be advantageous for the athlete, this protein might also enhance restoration of fluid balance.

Physical exercise increases metabolic rate and core body temperature and in an attempt to prevent a potentially dangerous rise in core temperature, heat loss mechanisms are initiated. During physical activity, particularly in warm environments, sweat rate is increased to facilitate this heat loss. Sweat secreted during exercise is generally hypotonic compared to plasma, which has the effect of raising plasma osmolality to stimulate drinking, but if insufficient drink is ingested to replace sweat losses, hypohydration (a state of negative water balance) will develop. Hypohydration by 2% body mass (~3% total body water) has consistently been shown to impair aerobic exercise performance in temperate, warm or hot conditions [1] and the greater the heat stress imposed, the greater the decrement in performance [2]. Although less consistent than the data supporting an impairment of aerobic exercise performance, there is also some evidence suggesting hypohydration might impair cognitive function [3] and sport-specific skills performance [4], as well as muscular strength and power [5].

It is commonly reported that athletes do not ingest enough drink during exercise to replace sweat losses and thus finish exercise hypohydrated [6]. Furthermore, it seems that many athletes commence competition [6] and training [7] in a hypohydrated state. In situations where there is adequate time (>24 h) between bouts of exercise for sufficient food and drink ingestion to allow complete recovery of body water stores, specific rehydration strategies are probably not warranted [1]. However, when the time between bouts of exercise is short or where athletes have intentionally dehydrated themselves as part of a rapid weight loss strategy [8], the use of a specific rehydration strategy to replace lost body water might be advantageous and prevent any decrement in subsequent exercise performance [9].

Post-Exercise Rehydration

There is now a substantial body of literature that has examined factors effecting rehydration after exercise and the main factors have been identified as the volume of drink ingested and the composition of the ingested drink [10]. It is clear that complete rehydration cannot occur without sufficient drink ingestion, but the drink volume ingested must actually be in excess of the volume of sweat lost to account for obligatory fluid losses that continue even in a hypohydrated state [11]. In situations where drinks are ingested ad libitum, palatability plays a key role in the rehydration process by ensuring sufficient drink is ingested to replace water losses [12], but if rapid complete recovery of fluid balance is required, drink volume should be prescribed. Whilst ingesting a sufficient volume of drink is vital in facilitating complete rehydration, it is the composition of the drink that determines how much of the drink is retained [11], although the rate of drinking appears to also play an important role [13]. Shirreffs et al. [11] reported that ingestion of a low sodium drink (23 mmol/l), even in a volume equivalent to 150 or 200% of sweat lost, did not fully restore fluid balance 6 h after drinking, but ingesting a high sodium drink (61 mmol/l) produced a positive fluid balance when the same volumes of drink were ingested (i.e. 150 and 200%).

As the major cation in the extracellular fluid, sodium is responsible for much of the extracellular fluid osmolality and plays a key role in water balance and the retention of drinks ingested after exercise [14]. After exercise-induced dehydration, the volume of a rehydration drink retained is directly related to the sodium content of the ingested drink [15–17] and for complete restoration of fluid balance, restoration of sodium balance is required [16]. Increasing the sodium content of a rehydration drink leads to a large proportion of this sodium being retained by the body [15, 16], attenuating the decline in serum sodium and osmolality that occurs following the ingestion of a large volume of a sodium-free drink and consequently reduces urine production [14].

As the major cation in the intracellular fluid, the addition of potassium to rehydration drinks has been suggested to increase post-exercise rehydration by increasing water retention in the intracellular fluid [10] and whilst one study has reported enhanced rehydration with potassium addition to a rehydration drink [18], another has reported no effect [19].

The effect of manipulating the carbohydrate concentration of a rehydration drink ingested after exercise has also been examined [20, 21] and it appears that the addition of relatively large amounts of carbohydrate (10–12% weight/volume) can reduce urine volume after drinking and produce a less negative net fluid balance compared to low carbohydrate/carbohydrate-free drinks. Increasing carbohydrate concentration also increases energy density, which is well known to reduce the rate at which a drink empties from the stomach, slowing water delivery to the intestine [22]. Furthermore, the hypertonic nature of these high-carbohydrate drinks leads to a net secretion of water from the circulation into the intestine to facilitate absorption [23], attenuating plasma volume restoration, offsetting the decline in serum osmolality and reducing urine production, at least transiently after drinking [20, 24]. Although these results should be interpreted with care as it is likely a large proportion of the water ingested will remain in the gastrointestinal tract for some time after drinking [24] and thus might not truly represent enhanced rehydration, at least not in the immediate post-drinking period.

From these studies it seems there are two mechanisms by which the composition of a drink might enhance rehydration. Firstly, via the addition of osmotically active substances that increase water retention in a particular body fluid compartment or secondly, by slowing the absorption of the drink via a reduction in gastric emptying and/or intestinal absorption. Both of these mechanisms attenuate the reduction in serum osmolality, which attenuates the decline in a number of fluid balance hormones, leading to increased renal water reabsorption and reduced urine production.

Milk Protein and Post-Exercise Rehydration

Whilst the main purpose of ingesting protein after exercise is to stimulate muscle protein synthesis and enhance muscle recovery [25], in recent years there has been some interest in the effects of milk [26, 27] and milk-derived proteins [28–30] on post-exercise rehydration. Shirreffs et al. [26] provided subjects water, a 6% carbohydrate-electrolyte sports drink, skimmed milk or skimmed milk plus an extra 20 mmol/l sodium chloride in a volume equivalent to 150% of the sweat lost during exercise. Four hours after drinking, subject's fluid balance was significantly negative following the water and sports drink trials, but was maintained in both milk trials. Using a similar design, Watson et al. [27] compared the rehydration effects of skimmed milk and a 6% carbohydrate-electrolyte sports drink, reporting that fluid balance was better maintained 3 h after drinking in the skimmed milk trial. Watson et al. [27] also examined exercise capacity in a hot environment, reporting that despite the difference in fluid balance, exercise capacity was similar for the skimmed milk (39.7 ± 8.1 min) and sports drink (39.6 ± 7.3 min) trials.

From the results of these two studies [26, 27] it is clear that skimmed milk enhances the restoration of fluid balance after exercise-induced dehydration compared to a carbohydrate-electrolyte sports drink, but the numerous compositional differences between skimmed milk and sports drinks make it difficult to determine what in the milk is responsible. Compared to sports drinks, skimmed milk contains additional sodium (10–15 mmol/l) and potassium (~40 mmol/l) [26, 27]. As previously discussed, drink sodium content is directly related to recovery of fluid balance, but the additional sodium in skimmed milk is unlikely to be sufficient to increase fluid balance to the extent observed [17]. Furthermore, the addition of 20 mmol/l extra sodium to skimmed milk had no effect on fluid balance [26]. Similarly, the success of potassium-containing rehydration drinks is equivocal [18, 19] and unlikely to account for the difference. Another compositional difference is that skimmed milk is more energy dense than sports drinks (1,480 vs. 1,020 kJ/l). As previously discussed, this is likely to reduce gastric emptying and the rate of absorption into the peripheral circulation, attenuating the decline in serum osmolality associated with ingesting large volumes of drink. In line with this, Watson et al. [27] observed that serum osmolality was greater after ingestion of skimmed milk compared to carbohydrate-electrolyte sports drink from 0 to 3 h after drinking. Although it is worth noting that the difference in energy density between the sports drink and water was much larger (1,020 vs. 0 kJ/l) and did not result in any difference in fluid balance restoration [26], suggesting this is unlikely to account for the enhanced restoration of fluid balance.

Whilst it cannot be discounted that some of the rehydration success of skimmed milk compared to sports drinks is simply the additive effects of these factors discussed above, it is possible that the protein content of milk or indeed the different carbohydrate type (lactose vs. glucose, maltodextrin or sucrose in sports drinks) accounts for these fluid balance effects. Whilst the effect of lactose on rehydration has not been investigated, the effect of milk protein or milk protein fractions has recently received some attention [28–30].

James et al. [29] examined the effect of adding complete milk protein to a 6.5% carbohydrate rehydration drink ingested in a volume equivalent to 150% of sweat losses. The milk protein was added in an energy-matched fashion such that some of the carbohydrate (2.5%) was replaced with an isoenergetic amount of milk protein. The addition of milk protein significantly enhanced the restoration of fluid balance 4 h after drinking, suggesting that gram-for-gram milk protein augments better rehydration than carbohydrate [29]. Similarly, Seifert et al. [28] observed that more of a 6% carbohydrate-1.5% whey protein drink was retained than either a 6% carbohydrate drink or a flavoured placebo. Seifert et al. [28] only provided a drink volume equivalent to sweat loss meaning subjects would never have been able to fully rehydrate [11]. More recently, James et al. [30] examined the effects whey protein on rehydration after exercise, providing subjects with 150% sweat losses of a 5% carbohydrate-1.5% whey protein isolate drink or an energy-matched 6.5% carbohydrate drink. Four hours after drinking, fluid balance was not different between the drinks, suggesting that when matched for energy density and provided in a volume great enough to facilitate complete restoration of fluid balance [11], whey protein does not appear to effect rehydration after exercise.

At present the mechanism behind the apparent difference in response between complete milk protein and whey protein is not known, but it might be related to the absorption kinetics of the different protein fractions. Milk protein is comprised of ~80% casein proteins and ~20% whey proteins and in the presence of gastric acid the casein fraction of milk protein clots [31], which might affect the rate at which a casein or milk protein-containing drink empties from the stomach. It has previously been demonstrated that gastric emptying (measured using a radioactive tracer) of drinks containing a high fraction of intact casein protein is slower than drinks containing a low fraction of casein protein or containing casein protein denatured by acidification [31]. Furthermore, gastric emptying of casein protein has been shown to be slower than glucose [32]. Other studies have reported that gastric emptying rate is linearly related to a drinks energy density and that the presence of milk protein has no additional effect [33] and that gastric emptying rate is not different between casein or whey protein-containing drinks [34]. These studies suggest that milk protein or casein protein have no effect on gastric emptying, but the findings might also be explained by the use of the double sampling gastric aspiration technique to measure gastric emptying rate. This involved repeated (at 5- to 20-min intervals) mixing of the stomach, via an orogastric/nasogastric tube, which might have disrupted any clot formation allowing faster gastric emptying.

Calbet and Holst [34] added tritium to drinks and observed a reduced tritium appearance in the peripheral circulation with a casein protein drink compared to an isoenergetic whey drink. As there was no difference in gastric emptying, this reduced tritium appearance indicates that the rate of intestinal water absorption was reduced in the casein protein-containing drink. As with high-concentration carbohydrate drinks [24], these potential effects of drinks containing intact casein protein on gastric emptying and/or intestinal absorption might impact on the retention of a drink containing milk protein by reducing the rate at which water enters the circulation and attenuating the reduction in serum osmolality after drinking [24].

Conclusions

From the results of these studies it can be concluded that ingesting milk or a carbohydrate-electrolyte rehydration drink with additional milk protein or whey protein is as effective as or more effective than a traditional carbohydrate-electrolyte rehydration drink at restoring fluid balance after exercise. This is of practical relevance to the training athlete whose post-exercise nutritional goals will be multifactorial and often rehydration will be necessary in combination with glycogen resynthesis and a stimulation of muscle protein synthesis. It is therefore beneficial to know that the addition of protein to a rehydration drink might enhance rehydration as well as providing amino acids to stimulate muscle protein synthesis [35] and possibly enhance muscle glycogen resynthesis in situations of suboptimal carbohydrate intake [36].

References

1 Sawka MN, Burke LM, Eichner ER, Maughan RJ, Montain SJ Stachenfeld NS: Exercise and fluid replacement. Med Sci Sports Exerc 2007;39:377–390.

2 Kenefick RW, Cheuvront SN, Palombo LJ, Ely BR, Sawka MN: Skin temperature modifies the impact of hypohydration on aerobic performance. J Appl Physiol 2010;109:79–86.

3 Ganio MS, Armstrong LE, Casa DJ, McDermott BP, Lee EC, Yamamoto LM, Marzano S, Lopez RM, Jimenez L, Le Bellego L, Chevillotte E, Lieberman HR: Mild dehydration impairs cognitive performance and mood of men. Br J Nutr 2011;7:1–9.

4 Baker LB, Dougherty KA, Chow M, Kenney WL: Progressive dehydration causes a progressive decline in basketball skill performance. Med Sci Sports Exerc 2009;39:1114–1123.

5 Judelson DA, Maresh CM, Anderson JM, Armstrong LE, Casa DJ, Kraemer WJ, Volek JS: Hydration and muscular performance: does fluid balance affect strength, power and high-intensity endurance? Sports Med 2007;47:907–921

6 Maughan RJ, Watson P, Evans GH, Broad N, Shirreffs SM: Water balance and salt losses in competitive football. Int J Sport Nutr Exerc Metab 2007;17:583–594.

7 Volpe SL, Poule KA, Bland EG: Estimation of pre-practice hydration status of National Collegiate Athletic Association Division I athletes. J Athl Train 2009;44:624–629.

8 Smith M: Physiological profile of senior and junior England international amateur boxers. J Sports Sci Med 2006;CSSI:74–89.

9 Judelson DA, Maresh CM, Farrell MJ, Yamamoto LM, Armstrong LE, Kraemer WJ, Volek JS, Spiering BA, Casa DJ, Anderson JM: Effect of hydration state on strength, power, and resistance exercise performance. Med Sci Sports Exerc 2007;39:1817–1824.

10 Shirreffs SM, Armstrong LE, Cheuvront SN: Fluid and electrolyte needs for preparation and recovery from training and competition. J Sports Sci 2004;22:57–63.

11 Shirreffs SM, Taylor AJ, Leiper JB, Maughan RJ: Post-exercise rehydration in man: effects of volume consumed and drink sodium content. Med Sci Sports Exerc 1996;28:1260–1271.

12 Wemple RD, Morocco TS, Mack GW: Influence of sodium replacement on fluid ingestion following exercise-induced dehydration. Int J Sport Nutr 1997;7:104–116.

13 Jones EJ, Bishop PA, Green JM, Richardson MT: Effects of metered versus bolus water consumption on urine production and rehydration. Int J Sport Nutr Exerc Metab 2010;20:139–144.

14 Nose H, Mack GW, Shi XR, Nadel ER: Role of osmolality and plasma volume during rehydration in humans. J Appl Physiol 1988;65:325–331.

15 Maughan RJ, Leiper JB: Sodium and post-exercise rehydration in man. Eur J Appl Physiol 1995;71:311–319.

16 Shirreffs SM, Maughan RJ: Volume repletion after exercise-induced volume depletion in humans: replacement of water and sodium losses. Am J Physiol 1998;274:F868–F875.

17 Merson SJ, Maughan RJ, Shirreffs SM: Rehydration with drinks differing in sodium concentration and recovery from moderate exercise-induced hypohydration in man. Eur J Appl Physiol 2008;103:585–594.

18 Maughan RJ, Owen JH, Shirreffs SM, Leiper JB: Post-exercise rehydration in man: effects of electrolyte addition to ingested fluids. Eur J Appl Physiol Occup Physiol 1994;69:209–215.

19 Shirreffs SM, Aragon-Vargas LF, Keil M, Love TD, Phillips S: Rehydration after exercise in the heat: a comparison of four commonly used drinks. Int J Sport Nutr Exerc Metab 2007;17:244–258.

20 Evans GH, Shirreffs SM, Maughan RJ: Post-exercise rehydration in man: the effects of osmolality and carbohydrate content of ingested drinks. Nutrition 2009;25:905–913.

21 Osterberg KL, Pallardy SE, Johnson RJ, Horswill CA: Carbohydrate exerts a mild influence on fluid retention following exercise-induced dehydration. J Appl Physiol 2010;108:245–250.

22 Vist GE, Maughan RJ: Gastric emptying of ingested solutions in man: effect of beverage glucose concentration. Med Sci Sports Exerc 1994;26:1269–1273.

23 Leiper JB, Maughan RJ: Absorption of water and electrolytes from hypotonic, isotonic and hypertonic solutions. J Physiol 1986;373:90P.

24 Evans GH, Shirreffs SM, Maughan RJ: The effects of repeated ingestion of high and low glucose-electrolyte solutions on gastric emptying and blood $2H_2O$ concentration after an overnight fast. Br J Nutr 2012;27:1–8.

25 Koopman R, Saris WH, Wagenmakers AJ, van Loon LJ: Nutritional interventions to promote post-exercise muscle protein synthesis. Sports Med 2007;37:895–906.

26 Shirreffs SM, Watson P, Maughan RJ: Milk as an effective post-exercise rehydration drink. Br J Nutr 2007;98:173–180.

27 Watson P, Love TD, Maughan RJ, Shirreffs SM: A comparison of the effects of milk and a carbohydrate-electrolyte drink on the restoration of fluid balance and exercise capacity in a hot, humid environment. Eur J Appl Physiol 2008;104:633–642.

28 Seifert J, Harmon J, DeClercq P: Protein added to a sports drink improves fluid retention. Int J Sport Nutr Exerc Metab 2006;16:420–429.

29 James LJ, Clayton D, Evans GE: Effect of milk protein addition to a carbohydrate-electrolyte rehydration solution ingested after exercise in the heat. Br J Nutr 2011;105:393–399.

30 James LJ, Gingell R, Evans GH: Effect of whey protein addition to a carbohydrate-electrolyte rehydration solution ingested after exercise in the heat. J Athl Train 2012;47:61–66.

31 Billeaud C, Guillet J, Sandler B: Gastric emptying in infants with or without gastro-oesophageal reflux according to the type of milk. Eur J Clin Nutr 1990;44:577–583.

32 Burn-Murdoch RA, Fisher MA, Hunt JN: The slowing of gastric emptying by protein in test meals. J Physiol 1978;274:477–485.

33 Calbert JAL, MacLean DA: Role of caloric content on gastric emptying in humans. J Physiol 1997;498.2:553–559.

34 Calbet JA, Holst JJ: Gastric emptying, gastric secretion and enterogastrone response after administration of milk proteins or their peptide hydrolysates in humans. Eur J Nutr 2005;43:127–139.

35 Howarth KR, Moreau NA, Phillips SM, Gibala MJ: Co-ingestion of protein with carbohydrate during recovery from endurance exercise stimulates skeletal muscle protein synthesis in humans. J Appl Physiol 2009;106:1394–1402.

36 Betts JA, Williams C: Short-term recovery from prolonged exercise: exploring the potential for protein ingestion to accentuate the benefits of carbohydrate supplements. Sports Med 2010;40:941–959.

Lewis James, PhD
School of Sport, Exercise and Health Sciences
Loughborough University
Loughborough LE11 3TU (UK)
Tel. +44 0 1509 226305, E-Mail L.James@Lboro.ac.uk

Lamprecht M (ed): Acute Topics in Sport Nutrition.
Med Sport Sci. Basel, Karger, 2013, vol 59, pp 127–134

Chocolate Milk: A Post-Exercise Recovery Beverage for Endurance Sports

Kelly Pritchett · Robert Pritchett

Department of Nutrition Exercise and Health Sciences, Central Washington University, Ellensburg, Wash., USA

Abstract

An optimal post-exercise nutrition regimen is fundamental for ensuring recovery. Therefore, research has aimed to examine post-exercise nutritional strategies for enhanced training stimuli. Chocolate milk has become an affordable recovery beverage for many athletes, taking the place of more expensive commercially available recovery beverages. Low-fat chocolate milk consists of a 4:1 carbohydrate:protein ratio (similar to many commercial recovery beverages) and provides fluids and sodium to aid in post-workout recovery. Consuming chocolate milk ($1.0–1.5 \cdot g \cdot kg^{-1} h^{-1}$) immediately after exercise and again at 2 h post-exercise appears to be optimal for exercise recovery and may attenuate indices of muscle damage. Future research should examine the optimal amount, timing, and frequency of ingestion of chocolate milk on post-exercise recovery measures including performance, indices of muscle damage, and muscle glycogen resynthesis.

Athletes, coaches, and sports dietitians have been striving to find a post-exercise nutritional strategy that will enhance muscle glycogen resynthesis, accelerate recovery, and maintain or improve the quality of future workouts or performances [1–4]. Athletes often participate in multiple training sessions a day with as little as 6 h of recovery between workouts [3]. A primary component of preparation for an elite athlete is to maximize their training in order to increase their potential for competition. Athletes are advised to adhere to nutritional strategies that maximize recovery [4], because full and often rapid recovery is vital for subsequent optimal performance [2]. Likewise, in some sports similar to swimming and track, athletes may compete multiple times in a single day. Enhanced recovery between competitions is important in these cases as well. Anything that improves recovery provides more effective successive practice sessions and should improve performance [3]. Furthermore, if athletes fail to take advantage of post-exercise nutritional strategies, recovery is impaired [4].

Recently, research has examined the efficacy of low-fat chocolate milk [5–12] on recovery measures. The literature examining the consumption of low-fat chocolate milk during recovery periods of ≥4 h suggests enhanced recovery [5–12], which would be beneficial to athletes competing in events with short recovery periods (preliminary heats, finals) such as track and field, swimming, and multiple day events, such as the stage, race cycling [4].

This chapter provides an overview and discussion of the literature that examines the efficacy of low-fat chocolate milk on various indices of recovery including muscle damage, glycogen resynthesis, protein synthesis and endurance exercise performance. Guidelines regarding the optimal timing and amount of chocolate milk post-exercise will also be discussed.

Low-Fat Chocolate Milk Composition

Low-fat chocolate milk has been suggested to be an effective, lower-cost recovery aid with a carbohydrate:protein (CHO:PRO) ratio similar to many commercial recovery and CHO-replacement beverages [5–12]. As previously mentioned, low-fat chocolate milk consists of cocoa plus monosaccharides (glucose and fructose) and disaccharides (lactose), and has a 3.5:1 CHO:PRO ratio similar to many commercial recovery beverages (CRBs). Being equivalent to a CRB is important because chocolate milk is well received by most athletes and is more economical and practical than most CRBs [4, 9]. In comparison to many CHO electrolyte beverages, chocolate milk boasts substantially more CHOs per milliliter. Crucial for rehydration, it also provides fluids and sodium [13], which need to be replaced due to sweat loss during a workout. Chocolate milk is also high in calcium, a major constituent in muscular contraction and necessary for building and maintaining strong bones. Based on the recommendations regarding post-exercise CHO intake [14], a 70-kg male would need to consume 510–810 ml (70–84 g CHO and 19–30 g PRO) and a 60-kg female 435–690 ml (60–72 g CHO and 16–26 g PRO) of low-fat chocolate milk per hour. These are reasonable amounts for most athletes [9].

Low-Fat Chocolate Milk and Exercise Recovery

Studies comparing chocolate milk to over-the-counter (OTC) recovery beverages have reported similar benefits [6, 9, 10] or improvements [5, 7, 11, 12] in exercise performance. Karp et al. [5] was the first study to examine the effectiveness of consuming chocolate milk as a post-exercise recovery aid between two endurance cycling sessions. After 4 h of recovery, cycling time to exhaustion at 70% of $Vo_{2\,max}$ was significantly ($p \leq 0.05$) longer (40.0 ± 14.7 min) for the chocolate milk trial compared to the OTC recovery beverage (26.8 ± 14.7 min) in trained cyclists [5]. The authors proposed that differences in performance demonstrated in the study may have been attributed to the different types of CHO in the beverages. Because increases in muscle glycogen levels during the early hours of recovery are greater with simple versus complex CHO [15], perhaps the 4-hour recovery period did not allow adequate time for the complete digestion of the complex CHOs in the CRB. The authors also speculated as to whether the higher fat content of chocolate milk may have increased the levels of free fatty acids in the blood, and possibly delaying glycogen depletion during the subsequent cycling trial to exhaustion and allowing subject to cycle longer [5]. This explanation seems plausible because, when post-exercise consumption of low-fat chocolate milk was compared to an isocaloric (matched for CHO) OTC recovery beverage (based on 1 g CHO•kg^{-1} of body weight/h post-exercise for the first 2 h) after a high-intensity fatiguing trial, the authors found no differences in cycling time to exhaustion at 85% of $Vo_{2\ max}$ [6]. Similarly, Spaccarotella et al. [10] found no significant differences in shuttle run performances between chocolate milk and a CHO electrolyte beverage.

Pritchett et al. [6] compared the post-exercise ingestion of low-fat chocolate milk to an OTC CRB matched for CHO and PRO content and found that the increase in creatine kinase (CK), a marker of muscle damage, was significantly reduced ($p < 0.05$) from pre- to post-exercise in the chocolate milk trial compared to the CRB trial. All participants (n = 10) preferred the taste of chocolate milk to the OTC recovery beverage.

Chocolate Milk and Muscle Glycogen

Muscle glycogen is the primary fuel source during high-intensity exercise and an important source during endurance exercise. Therefore, post-exercise glycogen restoration plays a very important role in the recovery process. Glycogen resynthesis is highly dependent on the extent of glycogen depletion, as well as the type, duration, and intensity of the exercise session [16]. For up to 6 h post-exercise, the rate of muscle glycogen resynthesis is accelerated compared to rest, and within 24 h post-exercise, complete restoration of glycogen stores can occur if sufficient amounts of CHO are consumed [1, 4, 17].

An increase in post-exercise glycogen synthase and exercise-induced increases in insulin sensitivity may be the potential mechanisms responsible for the 2- to 4-hour period of enhanced glycogen resynthesis following exercise. The literature suggests that low post-exercise muscle glycogen levels may provide a stimulus for this increase in glycogen synthase. Furthermore, the literature indicates that a recovery meal consumed within 2 h post-exercise, compared to no feeding, is more effective in improving recovery [18].

Because complete muscle glycogen resynthesis can take as long as 24 h, even under optimal conditions, studies have examined methods that increase the rate of muscle glycogen resynthesis [19]. Depending on the extent of glycogen depletion, consuming 1.0–1.5 g CHO•kg^{-1} h^{-1} immediately after exercise, and at 30-min intervals for up to 6 h post-exercise, appears to be optimal for adequate glycogen resynthesis. [5, 14, 20, 21]. On the other hand, if CHO intake is delayed by 2 h post-exercise, glycogen resynthesis rates have been found to be 45% lower [3, 22]. Furthermore, consumption of a CHO-only beverage, versus water or a placebo, has shown to be beneficial improving in endurance performance [4].

As previously mentioned, low-fat chocolate milk consists of monosaccharides (glucose and fructose) and disaccharides (lactose), while most commercially available recovery beverages consist of monosaccharides (glucose and fructose) and complex CHOs (maltodextrin) [4, 5]. Until recently, no studies had examined the effects of chocolate milk on muscle glycogen resynthesis post-exercise. Therefore, it was unclear as to whether these recovery benefits seen in the literature from chocolate milk can actually be attributed to the CHO:PRO ratio that has been suggested in the literature to enhance post-exercise recovery.

Lunn et al. [12] examined the efficacy of post-exercise fat-free chocolate milk consumption compared to a CHO-only beverage and determined that chocolate milk was as effective as the CHO-only beverage in maintaining muscle glycogen during the recovery period. Furthermore, the subsequent exercise performance was enhanced with chocolate milk when compared to the CHO-only beverage. Therefore, it appears that fat-free chocolate milk can be effective for maintaining muscle glycogen during the recovery period and possibly enhance subsequent performance. Further research should compare the effectiveness of chocolate milk to a CHO:PRO beverage on post-exercise muscle glycogen resynthesis. In addition, research is warranted to determine mechanisms of action behind the performance improvements observed in the literature.

Chocolate Milk and Protein Synthesis

Both high-intensity and prolonged endurance exercise can damage skeletal muscle resulting in delayed-onset muscle soreness with concurrent increases in markers of muscle damage such as CK, myoglobin, cortisol, and lactate dehydrogenase [23, 24]. Elevated levels of these enzymatic markers are associated with decreased performance [24].

The addition of PRO to a post-exercise recovery meal may enhance net PRO anabolism [25]. During the post-exercise period, there is an increased rate of muscle PRO synthesis in trained individuals [3]. Research examining subsequent exercise performance and muscle damage suggests that a CHO:PRO versus a CHO-only beverage ingested post-exercise may enhance recovery [4, 25–27].

Some studies show improved performance with CHO-PRO post-exercise complex versus a CHO-only [5, 25, 27–30], while others show no difference in performance [31–34]. It should be noted that the majority of the literature has examined recovery in trained cyclists or runners when performing a time trial to exhaustion at 70–85% of $Vo_{2 \, max}$ [5, 25, 27–29, 31–34].

Furthermore, studies have reported decreases in muscle damage (CPK) with the addition of PRO to a recovery beverage after exercise sessions [6, 18, 21, 25, 27, 35]. Due to the applied nature of recovery studies, the majority of the literature examining muscle damage has included multiple indicators of muscle damage including blood-borne CK, and subjective measures of muscle soreness (using a visual scale) [36]. However, CK has been criticized as an effective indicator of muscle damage because of poor correlations with indices of muscle damage [4].

Post-exercise recovery beverages containing PRO seem to be effective in improving recovery indices. However, some of the results may be due to the higher caloric content of the CHO:PRO supplements. The additional PRO calories via glucogenesis may have provided additional substrate for glycogen resynthesis to occur, therefore aiding in an enhanced recovery [22]. Currently, research suggests that 20–25 g of high-quality PRO during a single feeding is optimal [2, 4].

Low-fat chocolate milk is a high-quality PRO containing all nine essential amino acids and has a 4:1 CHO:PRO ratio similar to CRBs. Although there are numerous CHO:PRO supplements on the market, recent studies have examined the efficacy of chocolate milk on PRO synthesis following endurance exercise [11, 12]. The results of these studies suggest that chocolate milk is more effective than an isocaloric CHO-only beverage in creating an anabolic intracellular environment post-exercise [11, 12]. While, further research is warranted to examine mechanisms of action, current research suggests that chocolate milk is a viable alternative to support post-exercise PRO synthesis [11].

Chronic Supplementation with Chocolate Milk

Research has also reported improved performance and improvements in recovery indices, as indicated by elevated CK, with low-fat chocolate milk when given during and after the exercise session (CITE). However, the majority of the studies have examined the effects of a single dose, post-exercise beverage on muscle damage and recovery indices [9]. Very few studies [8, 9] have examined the effects of chocolate milk taken over time (~1 week) on indices of recovery. For practicality purposes, it is beneficial to examine the effects of a beverage over time, because an ideal beverage supplement will benefit daily workouts.

Gilson et al. [8] examined the effectiveness of low-fat chocolate milk versus a high CHO recovery beverage consumed post-exercise for a week in collegiate soccer players. The soccer players continued their normal training regimen, which was similar among subjects. This study found significantly ($p \leq 0.05$) lower CK (chocolate milk 316.9 ± 188.3 U•l^{-1}, CHO 431.6 ± 310.8 U•l^{-1}) levels after 1 week of supplementation with chocolate milk versus a high CHO-only beverage. However, no performance differences were detected between the two beverages. Similarly, Pritchett et al. [9] found no significant difference between low-fat chocolate milk and an OTC recovery beverage, matched for CHO and PRO content, in time to exhaustion at 85% of $Vo_{2\,peak}$ or markers of muscle damage (CK) after a week of post-exercise supplementation. Based on the current literature, low-fat chocolate milk results in similar performance outcomes to an OTC recovery beverage when ingested over a week-long period. Future research should examine the effects of chocolate milk taken over a longer period of time.

Conclusions

The available literature regarding post-exercise nutritional strategies for optimal performance is constantly evolving. The optimal timing regarding post-exercise nutritional strategies for maximal glycogen resynthesis appears to be within the first 2 h post-exercise [16, 28, 37, 38]. The literature suggests that consuming 1.0–1.5 $g\bullet kg^{-1}$ h^{-1} for the first 2 h post-exercise of chocolate milk may be optimal for recovery [5–8, 10, 39]. The additional PRO (20–25 g) in the food or beverage may aid in muscle PRO resynthesis [2, 16, 28, 36, 38]. Recent literature supports the efficacy of chocolate milk as a recovery aid between workouts [5–12, 39]. Low-fat chocolate milk serves as a more convenient, cheaper, premixed, and, based on current observations, more palatable, recovery beverage option for many athletes. However, given the limited number of studies, further research is still warranted to examine the efficacy of chocolate milk on exercise recovery. Further research is needed to examine the dosage, timing and frequency of chocolate milk ingestion on recovery indices including performance, muscle glycogen resynthesis, muscle soreness and oxidative stress levels [4, 9].

References

1 Jentjens RL, Jeukendrup AE: Determinants of post-exercise glycogen synthesis during short-term recovery. Int J Sport Nutr Exerc Metab 2003;33:117–144.

2 Burke L: Fasting and recovery from exercise. Br J Sports Med 2010;44:502–508.

3 Bishop PA, Jones E, Woods K: Recovery from training: a brief review. J Strength Cond Res 2008; 229:1–10.

4 Pritchett K, Pritchett R, Bishop P: Nutritional strategies for post-exercise recovery: a review. S Afr J Sports Med 2011;23:20–25.

5 Karp JR, Johnston JD, Tecklenburg S, Mickleborough TD, Fly AD, Stager JM: Chocolate milk as a post-exercise recovery aid. Int J Sport Nutr Exerc Metab 2006;16:78–91.

6 Pritchett KL, Bishop PA, Pritchett RC, Green JM, Katica C: Acute effects of chocolate milk and a commercial recovery beverage on post-exercise muscle damage and cycling performance. J Appl Phys Nutr Metab 2009;34:1017–1022.

7 Thomas K, Morris P, Stevenson E: Improved endurance capacity following chocolate milk consumption compared with two commercially available sports drinks. Appl Physiol Nutr Metab 2009;34:78–82.

8 Gilson S, Saunders MJ, Moran C, Moore R, Womack CJ, Todd K: Effects of chocolate milk consumption on markers of recovery following soccer training: a randomized cross-over study. J Int Soc Sports Nutr 2010;7:19.

Pritchett · Pritchett

9 Pritchett KL, Bishop PA, Pritchett RP, Green JM, Combs B, Eldridge M, Katica C: Comparisons of post-exercise chocolate milk and a commercial recovery beverage following cycling training on recovery and performance. J Exerc Phys Online 2011;14:29–39.

10 Spaccarotellaa KJ, Andzel WD: The effects of low fat chocolate milk on post-exercise recovery in collegiate athletes. J Strength Cond Res 2011;25:3456–3460.

11 Ferguson-Stegall L, McCleave EL, Ding M, Doerner PG III, Wang B, Liao Y, Liu Y, Hwang J, Dessard BM, Ivy JL: Post-exercise carbohydrate-protein supplementation improves subsequent exercise performance and intracellular signaling for protein synthesis. J Strength Cond Res 2011;25:12–24.

12 Lunn WR, Pasiakos SM, Colletto MR, Karfonta KE, Carbone JW, Anderson JM, Rodriguez NR: Chocolate milk and endurance exercise recovery: protein balance, glycogen, and performance. Med Sci Sports Exerc 2012;44:682–690.

13 Clapp AJ, Bishop PA, Smith JF, Mansfield ER: Effects of carbohydrate-electrolyte content of beverages on voluntary hydration in a simulated industrial environment. Am Ind Hyg Assoc J 2000;61:692–699.

14 American College of Sports Medicine, American Dietetic Association, and Dietitians of Canada. Nutrition and Athletic Performance. Joint Position Statement of the American Dietetic Association, Dietitians of Canada, and the Medicine and American College of Sports Medicine. Med Sci Sports Exerc 2009;109:509–527.

15 Freidman JE, Neufer PD, Dohm GL: Regulation of glycogen resynthesis following exercise. Sports Med 1991;11:232–243.

16 Van Loon LJ, Saris WM, Ruijshoopanda MK, Wagenmakers AM: Maximizing post-exercise muscle glycogen synthesis: carbohydrate supplementation and the application of amino acid or protein hydrolysate. Am J Clin Nutr 2000;72:106–111.

17 Ryans M: Sports Nutr Endur Athletes, ed 2. Boulder/CO, VeloPress, 2007.

18 Romano-Ely BC, Todd K, Saunders MJ, St Laurent T: Effect of an isocaloric carbohydrate-protein-antioxidant drink on cycling performance. Med Sci Sports Exerc 2006;38:1608–1616.

19 Van Hall G, Shirreffs SM, Calbert JAL: Muscle glycogen resynthesis during recovery from cycle exercise: no effect of additional protein ingestion. J Appl Physiol 2000;88:1631–1636.

20 Doyle JA, Sherman WM, Strauss RL: Effects of eccentric and concentric exercise on muscle glycogen replenishment. J Appl Physiol 1993;74:1848–1855.

21 Saunders MJ, Luden ND, Herrcick JE: Consumption of an oral carbohydrate-protein gel improves cycling endurance and prevents postexercise muscle damage. J Strength Cond Res 2007;21:678–684.

22 Ivy JL, Katz AL, Cutler CL: Muscle glycogen synthesis after exercise: effect of time of carbohydrate ingestion. J Appl Physiol 1988;64:1480–1485.

23 Clarkson PM, Kearns AK, Rouzier P, Rubin R, Thompson PD: Serum creatine kinase levels and renal function measures in exertional muscle damage. Med Sci Sports Exerc 2006;38:623–627.

24 White JP, Wilson JM, Austin KG, Greer BK, St John N, Panton LB: Effect of a carbohydrate-protein supplement timing on acute exercise-induced muscle damage. J Int Soc Sports Nutr 2008;5:5.

25 Saunders MJ, Kane MD, Todd MK: Effects of carbohydrate-protein beverage on cycling endurance and muscle damage. Med Sci Sport Exerc 2004;36:1233–1238.

26 Skillen RA, Testa M, Applegate EA, Heiden EA, Fascetti AJ, Casazza GA: Effects of an amino acid-carbohydrate drink on exercise performance after consecutive-day exercise bouts. Int J Sport Nutr Exerc Metab 2008;18:473–492.

27 Valentine RJ, Saunders MJ, Todd MK, St Laurent TG: Influence of carbohydrate-protein beverage on cycling endurance and indices of muscle disruption. Int J Sports Nutr Exerc Metab 2008;18:379–388.

28 Williams MB, Raven PB, Fogt DL, Ivy JL: Effects of recovery beverages on glycogen restoration and endurance exercise performance. J Strength Cond Res 2003;17:12–19.

29 Ivy JL, Res PT, Sprague RC, Widzer MO: Effect of a carbohydrate-protein supplement on endurance performance during exercise of varying intensity. Int J Sport Nutr Exerc Metab 2003;13:382–395.

30 Berardi JM, Noreen EE, Lemon PWR: Recovery from a cycling time trial is enhanced with carbohydrate-protein supplementation vs. isoenergetic carbohydrate supplementation. J Int Soc Sports Nutr 2008;5:24.

31 Berardi JM, Price TB, Noreen EE, Lemon PWR: Postexercise muscle glycogen recovery enhanced with carbohydrate-protein supplement. Med Sci Sports Exerc 2006;38:1106–1113.

32 Betts JA, Stevenson E, Williams C, Steppard C, Grey E, Griffin J: Recovery of endurance running capacity: effect of carbohydrate-protein mixtures. Int J Sport Nutr Exerc Metab 2005;15:590–609.

33 Carrithers JA, Williamson DL, Gallagher PM, Godard MP, Schulze KE, Trappe SW: Effects of postexercise carbohydrate-protein feedings on muscle glycogen restoration. J Appl Physiol 2000;88:1976–1982.

34 Green MS, Corona BT, Doyle JA, Ingalls CP: Carbohydrate-protein drinks do not enhance recovery from exercise-induced muscle injury. Int J Sports Nutr Exerc Metab 2008;18:1–18.

35 Ready SL, Seifert JL, Burke E: The effects of two sports drinks formulations on muscle stress and performance. Med Sci Sports Exerc 1999;31:S119.

36 Luden ND, Saunders MJ, Todd K: Postexercise carbohydrate-protein-antioxidant ingestion decreases plasma creatine kinase and muscle soreness. Int J Sport Nutr Exerc Metab 2007;17:109–123.

37 Zawadzki KM, Yaspelkis BB, Ivy JL: Carbohydrate-protein complex increases the rate of muscle glycogen storage after exercise. J Appl Physiol 1992;72:1854–1859.

38 Ivy JL, Goforth HW, Damon BM, McCauley TR, Parsons EC, Price TB: Early postexercise muscle glycogen recovery is enhanced with a carbohydrate-protein supplement. J Appl Physiol 2002;93:1337–1344.

39 Spaccarotellaa KJ, Andzel WD: Building a beverage for recovery from endurance activity: A review. J Strength Cond Res 2011;25:3198–3204.

Kelly Pritchett, PhD, RD, CSSD
Department of NEHS, Central Washington University
PE Building, Room 137, 400 E. University Way
Ellensburg, WA 98926-7552 (USA)
Tel. +1 509 963 2351, E-Mail kkerr@cwu.edu

Lamprecht M (ed): Acute Topics in Sport Nutrition.
Med Sport Sci. Basel, Karger, 2013, vol 59, pp 135–142

Role of Supplementary L-Carnitine in Exercise and Exercise Recovery

Amy Huang · Kevin Owen

Lonza Ltd, Basel, Switzerland

Abstract

L-Carnitine is a conditionally essential nutrient and plays an important role in mitochondrial β-oxidation. As a dietary supplement for athletes, L-carnitine has been investigated for its potential to enhance β-oxidation during exercise ultimately to improve performance. While some studies have shown a positive impact on $Vo_{2\,max}$ and other performance measures, other studies have found contradictory results. As such, investigations to a different mechanism by which L-carnitine supplementation could impact exercise and recovery were explored. Based on findings from cardiovascular research that L-carnitine enhances vascular endothelial function, an alternate hypothesis was developed. The hypothesis is centered on improving blood flow to muscle tissues and decreasing hypoxic stress and its resulting sequelae. Studies have shown a decrease in markers of purine catabolism and free radical generation and muscle soreness as a result of L-carnitine supplementation. Direct assessment of muscle tissue damage via magnetic resonance imaging also indicates the ability of L-carnitine to attenuate tissue damage related to hypoxic stress. L-Carnitine is regarded as a safe supplement for athletes and has been shown to positively impact the recovery process after exercise.

L-Carnitine was first isolated in meat extracts in the early 1900s. Later in the 1950s, the metabolic function of this new compound was realized. In the years following, extensive research was conducted to help understand the biological function and metabolic role L-carnitine has in the body. This research led to the characterization of the primary function of L-carnitine as a transport molecule to facilitate fatty acid oxidation. Given this role of L-carnitine, a great deal of research has examined the potential of L-carnitine supplementation to enhance lipid oxidation, spare muscle glycogen, and improve exercise performance. The purpose of this review is to summarize the current literature regarding L-carnitine and its potential role in sports and exercise recovery.

Biosynthesis and Dietary Intake

L-Carnitine, or 3-hydroxy-3-N,N,N-trimethylaminobutyrate, is a quaternary amine and conditionally essential nutrient. It is biosynthesized in the presence of lysine and methionine and requires sufficient iron, vitamin C, vitamin B_6 and niacin [1]. The pathway also requires specific enzymes for complete biosynthesis from protein-bound lysine to L-carnitine. One such enzyme, γ-butyrobetaine hydroxylase, is highly expressed in the liver, testes and kidney, but not in cardiac or skeletal muscles. Therefore, endogenously synthesized circulating L-carnitine is largely from hepatic or renal origins. In conditions where the essential amino acids or cofactors are not present in sufficient quantities, it becomes essential to achieve adequate L-carnitine status through dietary intake. L-Carnitine is found primarily in foods of animal origin and only in limited quantities in plant foods. Sources of L-carnitine include beef steak, ground beef, pork and pork products, and chicken [2, 3]. It is estimated that 75% of L-carnitine in the body is attained from dietary intake [1].

Functions of L-Carnitine

L-Carnitine is stored primarily in heart and skeletal muscles and functions as a transporter of long-chain fatty acids across the mitochondrial membrane for β-oxidation. The rate of β-oxidation is regulated by carnitine palmitoyl transferase 1 (CPT1) and it has been shown that L-carnitine supplementation can increase CPT1 activity [4, 5]. Moreover, two separate, independent studies have demonstrated that L-carnitine supplementation can increase β-oxidation in healthy human subjects [6, 7]. L-Carnitine also modulates the ratio of free coenzyme A to acetyl-CoA. Free L-carnitine combines with acetyl-CoA in the mitochondria to form acetyl-L-carnitine groups, which results in an increase in mitochondrial free coenzyme A. Free coenzyme A can be utilized for ATP generation through intermediary metabolic pathways, such as the TCA (tricarboxylic acid) cycle [8].

It is well established that endurance exercise requires the use of long-chain fatty acids as an energy substrate. Early research in L-carnitine and exercise focused on the potential to enhance exercise performance by increasing fatty acid oxidation in skeletal muscle, thereby reducing muscle glycogen usage and increasing overall exercise capacity. Another hypothesis by which L-carnitine may improve exercise performance lies in the second metabolic function discussed above. When L-carnitine acts as an acetyl-CoA buffer, negative feedback from excess acetyl-CoA is decreased and ATP generation through glycolysis and the TCA cycle can continue. Moreover, buffering of acetyl-CoA may prevent the formation of muscle lactate, thereby improving exercise performance [9]. The underlying assumption behind these hypotheses was that muscle L-carnitine content could be increased through dietary supplementation.

Since cardiac and skeletal muscles are unable to synthesize L-carnitine due to a lack of γ-butyrobetaine hydroxylase, it must be transported from the plasma into muscle cells. While studies have shown an increase in plasma L-carnitine after supplementation, few studies have shown subsequent increases in muscle L-carnitine content [10, 11]. Wall et al. [9] have effectively increased muscle L-carnitine content during long-term L-carnitine supplementation in hyperinsulinemic states, however more research should be conducted in this area to elucidate the mechanism of action by which this is possible.

Consistent with the inability to increase muscle L-carnitine content, there are incongruous results with L-carnitine supplementation and exercise performance. While L-carnitine supplementation has been shown to improve exercise tolerance and activity levels in patients with cardiovascular disease [12–14], there are inconsistent results in healthy human subjects. Swart et al. [15] measured exercise performance in marathon runners after 6 weeks of L-carnitine supplementation and found a positive impact on peak treadmill running speed as well as $Vo_{2\,max}$. Vecchiet et al. [16] also found that supplementation with L-carnitine significantly increased $Vo_{2\,max}$ as well as power output. On the other hand, Krähenbühl [17] measured the impact of 3 months of L-carnitine supplementation on physical performance and found no impact on performance measures.

The inconsistency in the results regarding L-carnitine and exercise performance led way to shifting the focus of research towards exploring a different mechanism by which L-carnitine could impact exercise and recovery. Dubelaar et al. [18] demonstrated the impact of L-carnitine administration on muscle contractile force in dogs. Contractile force increased by 30% in L-carnitine-supplemented animals and blood flow was also significantly increased. Moreover, there was an increase in plasma L-carnitine levels while there was no increase in muscle L-carnitine content. From these results, the theory that a non-muscle mechanism was responsible for the increase in contractile force and improved blood flow emerged. Dubelaar et al. hypothesized that L-carnitine was acting on vascular cells surrounding the muscle and increasing oxygenation of the musculature. This hypothesis was based on the prior work of Hülsmann and Dubelaar [19], who had previously found that vascular endothelial cells preferentially use long-chain fatty acids as a source of energy. In later work, Hülsmann and Dubelaar [20] found that vascular endothelial cells may become deficient of L-carnitine during ischemia, thereby leading to a disruption in capillary blood flow.

Findings by Giamberadino et al. [21] also support the notion that L-carnitine may impact exercise through an alternative mechanism. When investigating the impact of L-carnitine supplementation on DOMS (delayed-onset muscle soreness) in healthy, untrained men, L-carnitine supplementation at 3 g/day for 3 weeks significantly reduced creatine kinase and pain as compared with placebo. The investigators attributed the findings to a vasodilatory effect from L-carnitine which was hypothesized to reduce hypoxic stress. This was the first study in humans to investigate the potential use of L-carnitine in recovery, and provided a basis for further research.

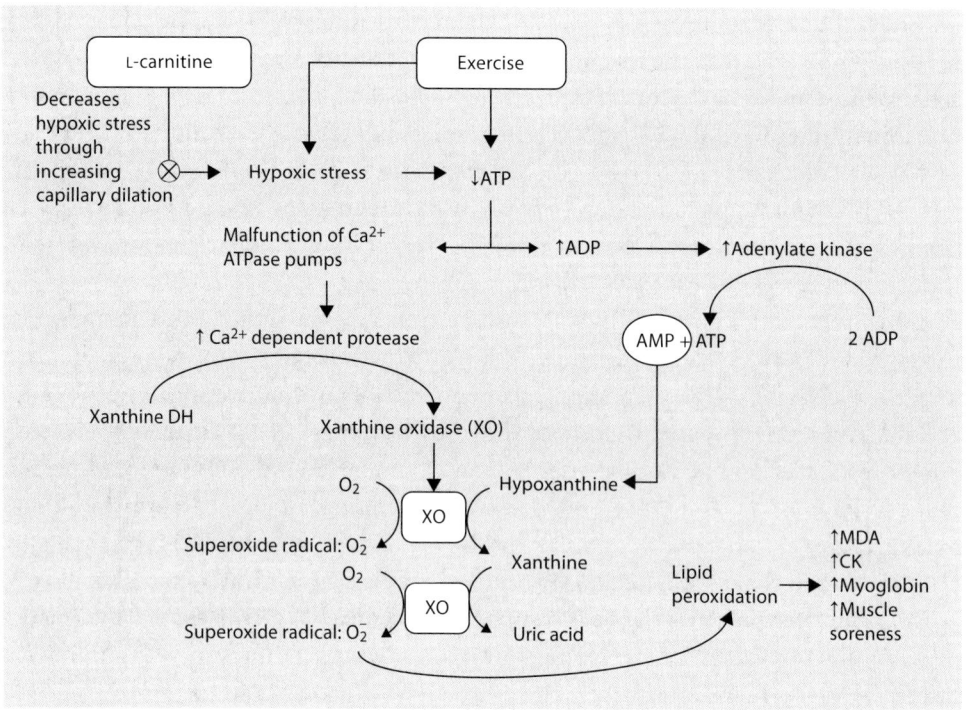

Fig. 1. Schematic diagram of L-carnitine's impact on exercise recovery. Adapted from Kraemer et al. [23].

Researchers from the Human Performance Laboratory from the University of Connecticut have expanded on the hypothesis that L-carnitine, from Carnipure™ tartrate, can impact the recovery process after exercise. It is hypothesized that L-carnitine is present in capillary sphincter beds and supplementation prevents depletion of L-carnitine stores [22]. By ensuring adequate supply of L-carnitine, vascular endothelial cells can metabolize long-chain fatty acids for energy, which protects the integrity of the vasculature surrounding the muscles. When there is increased blood flow to muscle tissues, hypoxia associated with exercise is mitigated, as is the resulting cascade of biochemical events, such release of enzymatic triggers which increase free radical formation (fig. 1). As a result, structural damage to muscle tissue may be decreased, allowing for an increase in intact androgen receptors, which facilitate protein synthesis and exercise recovery [23].

To further test the hypothesis that L-carnitine supplementation may help improve blood flow, Volek et al. [24] investigated the impact of L-carnitine supplementation on flow-mediated dilation in the brachial artery after a high fat meal. High fat meals are known stressors to vascular endothelial cells and thus an appropriate model to test the impact of L-carnitine on blood flow. The authors found that there were significant differences in postprandial flow-mediated dilation in the L-carnitine supplement

Huang · Owen

group as compared with placebo. This study further indicates that L-carnitine plays a beneficial role in mediating blood flow in response to a stressor on vascular endothelial cells.

As a result of improved blood flow, muscle cell disruption due to hypoxia is diminished, as evidenced by a decrease in indirect biochemical markers, such as myoglobin and hypoxanthine. Multiple human intervention studies have found that L-carnitine supplementation decreases biochemical markers of hypoxic stress after exercise. A study by the Human Performance Laboratory at the University of Connecticut examined the impact of L-carnitine supplementation on biochemical markers of exercise stress. Ten resistance-trained men participated in the crossover trial in which the volunteers consumed a placebo or L-carnitine for 3 weeks prior to performing a squat exercise. Markers of purine catabolism were measured, including hypoxanthine, xanthine oxidase, and uric acid. Circulating cytosolic proteins, such as myoglobin, fatty acid-binding protein and creatine kinase, were also measured. It was found that these biochemical markers of stress were significantly decreased after L-carnitine supplementation [25].

This effect has been replicated in several other studies, including one by Ho et al. [26, 27] which included a middle-aged, untrained population. Ho et al. employed a similar study design but included older, recreationally active adults in order to test the hypothesis that the same effects could be seen in this population. Men and women in the crossover study ingested 2 g of L-carnitine or placebo per day for 3 weeks and then performed a series of squats or leg presses. As was seen in the aforementioned study, L-carnitine attenuated the rise of the markers of purine catabolism as well as malondialdehyde, a marker of free radical formation. The results of this study indicate that L-carnitine supplementation may have beneficial impacts in individuals across the age and training spectrums.

It is clear that L-carnitine plays a role in mitigating markers of hypoxic stress, of which can lead to muscle damage. It is also interesting to note that researchers have directly assessed the impact of L-carnitine supplementation on muscle damage through the use of magnetic resonance imaging [25, 28]. Kraemer et al. [28] found that muscle damage, measured as a cross section of the mid-thigh, at post-exercise days 3 and 6 was 23 ± 8 and $16 \pm 5\%$ for the L-carnitine group and 39 ± 5 and $29 \pm 6\%$ for placebo, respectively. The percentage of muscle disruption was significantly lower for the L-carnitine group. The results of this study indicate that L-carnitine supplementation can impact overall tissue damage, as evidenced by direct imaging assessment.

Another benefit L-carnitine may confer in exercise recovery relates to androgenic responses. Androgens, such as testosterone, are key hormones for muscle rebuilding and repair after exercise. Androgens bind with androgen receptors on muscle cells which then initiate the cascade for transcription and translation to muscle protein. Inoue et al. [29] showed that obstruction of the androgen receptor negatively impacted skeletal muscle hypertrophy. Given that androgen receptors are integral for the androgen response, Kraemer et al. [30] hypothesized that L-carnitine could increase the androgen response due to its mitigating effects on hypoxic stress, thereby increasing the availability of androgen receptors due to less muscle damage. In a randomized, double-blind, placebo-controlled crossover trial to test this hypothesis, the investigators found that L-carnitine supplementation upregulated pre-exercise androgen receptor content as compared with placebo. Moreover, post-exercise supplementation with L-carnitine also significantly increased androgen receptor content and slightly decreased total testosterone. Such a decrease in testosterone suggests an increase in bindings to androgen receptors which could lead to enhanced protein synthesis and improved recovery after exercise. Although the impact was significant, it is important to note that dietary modification may also confer similar benefits for exercise recovery.

The effect of L-carnitine on exercise recovery has also been assessed by subjective measures. In the investigation by Ho et al. [26], participants rated their perceived muscle soreness using a visual analogue scale. Both men and women reported lower muscle soreness scores during L-carnitine supplementation as compared with placebo both during the exercise period and during the recovery days. This clearly shows that L-carnitine mitigated muscle soreness, likely due to decreased muscle tissue damage mediated by improved blood flow from L-carnitine supplementation.

As with any dietary supplement for athletes and the general population, safety and dosage must be considered. Spiering et al. [31] investigated the dose response of L-carnitine in athletes in order to elucidate the minimum effective dose needed for exercise recovery. The authors found that both 1- and 2-gram daily doses of L-carnitine were effective in mitigating biochemical markers of hypoxic stress as well as muscle soreness. In addition to dosage, the safety of L-carnitine supplementation has been extensively studied. In a study by Rubin et al. [32], 3 g of L-carnitine per day for 3 weeks was given to study volunteers. A plethora of clinical markers were assessed, including markers for hepatic and renal function as well as electrolytes and hematological parameters. The authors found no adverse effects on the measured parameters and concluded that L-carnitine is safe when used as a dietary supplement. The Scientific Committee on Food from the European Food Safety Authority has also evaluated the safety of L-carnitine tartrate. L-Carnitine tartrate is a non-hygroscopic salt that is pH- and heat-stable to 180°C. The panel concluded that the use of L-carnitine tartrate in foods in a daily dose of 3 g, which is equivalent to 2 g L-carnitine, poses no safety risk to consumers [33].

Conclusion

L-Carnitine is unique in its essential role in energy metabolism, transporting fatty acids across the mitochondrial membrane for subsequent breakdown and energy generation. Over the past 10 years a novel paradigm has emerged with research utilizing

L-carnitine. This paradigm places L-carnitine in the important role of facilitating the recovery process in response to a hypoxic stimulus such as physical activity. In this role, L-carnitine helps to protect the endothelial cells from an L-carnitine deficiency, mediate the markers of purine catabolism, reduce tissue damage and muscle soreness, and facilitate the overall process of recovery after exercise. Therefore, L-carnitine is an emerging supplement that may well have targeted and specific roles to play in the exercise domain as well as in other domains involving clinical populations.

Disclosure Statement

At the time of manuscript preparation, the author and co-author were employees of Lonza Ltd, which manufactures the ingredient Carnipure™, a source of L-carnitine.

References

1 Flanagan JL, Simmons PA, Vehige J, Willcox MDP, Garrett Q: Role of carnitine in disease. Nutr Metab (Lond) 2010;7:30–44.

2 Krähenbuhl S: L-Carnitine and vegetarianism. Ann Nutr Metab 2000;44:75–96.

3 Knüttel-Gustavesen S, Harmeyer J: The determination of L-carnitine in several food samples. Food Chem 2007;105:793–804.

4 Schreurs M, Kuipers F, van der Leij FR: Regulatory enzymes of mitochondrial β-oxidation as targets for treatment of the metabolic syndrome. Obes Rev 2010;11:380–388.

5 Karlic H, Lohninger S, Koeck T, Lohninger A: Dietary L-carnitine stimulates carnitine acyltransferases in the liver of aged rats. J Histochem Cytochem 2002;50:205–212.

6 Müller DM, Seim H, Kiess W, Löster H, Richter T: Effects of oral L-carnitine supplementation on in vivo long-chain fatty acid oxidation in healthy adults. Metabolism 2002;51:1389–1391.

7 Wutzke KD, Lorenz H: The effect of L-carnitine on fat oxidation, protein turnover, and body composition in slightly overweight subjects. Metabolism 2004;53:1002–1006.

8 Stephens FB, Constantin-Teodosiu D, Greenhaff PL: New insights concerning the role of carnitine in the regulation of fuel metabolism in skeletal muscle. J Physiol 2007;581.2:431–444.

9 Wall BT, Stephens FB, Constantin-Teodosiu D, Marimuthu K, Macdonald IA, Greenhaff PL: Chronic oral ingestion of L-carnitine and carbohydrate increases muscle carnitine content and alters muscle fuel metabolism during exercise in humans. J Physiol 2011;589.4:963–973.

10 Barnett C, Costill DL, Vukovich MD, Cole KJ, Goodpaster BH, Trappe SW, Fink WJ: Effect of L-carnitine supplementation on muscle and blood carnitine content and lactate accumulation during high-intensity sprint cycling. Int J Sport Nutr 1994; 4:280–288.

11 Vukovich MD, Costill DL, Fink WJ: Carnitine supplementation: effect on muscle carnitine and glycogen content during exercise. Med Sci Sports Exerc 1994;26:1122–1129.

12 Iyer RN, Khan AA, Gupta A, Vajifdar BU, Lokhandwala YY: L-Carnitine moderately improves the exercise tolerance in chronic stable angina. J Assoc Physicians India 2000;48:1050–1052.

13 Cacciatore L, Cerio R, Ciarimboli M, Cocozza M, Coto V, D'Alessandro A, D'Alessandro L, Grattarola G, Imparato L, Lingetti M, Mancini M, Oliviero U, Policicchio D, Porfido FA, Rengo F, Sorrentino GP: The therapeutic effect of L-carnitine in patients with exercise-induced stable angina: a controlled study. Drugs Exp Clin Res 1991;17:225–235.

14 Cherchi A, Lai C, Angelino F, Trucco G, Caponnetto S, Mereto PE, Rosolen G, Manzoli U, Schiavoni G, Reale A, Romeo F, Rizzon P, Sorgente L, Strano A, Novo S, Immordino R: Effects of L-carnitine on exercise tolerance in chronic stable angina: a multicenter, double-blind, randomized placebo controlled crossover study. Int J Clin Pharm Ther Toxicol 1985;23:569–572.

15 Swart I, Rossouw J, Loots JM, Kruger MC: The effect of L-carnitine supplementation on plasma carnitine levels and various performance parameters of male marathon athletes. Nutr Res 1997;17: 405–414.

16 Vecchiet L, DiLisa F, Pieralisi G, Ripari P, Menabo R, Giamberadino MA, Siliprandi N: Influence of L-carnitine administration on maximal physical exercise. Eur J Appl Physiol 1990;61:486–490.

17 Krähenbuhl S: L-Carnitine and physical performance. Ann Nutr Metab 2000;44:75–96.

18 Dubelaar M-L, Lucas C, Hülsman WC: The effect of L-carnitine on force development of the latissimus dorsi muscle in dogs. J Cardiac Surg 1991;6:270–275.

19 Hülsmann WC, Dubelaar ML: Aspects of fatty acid metabolism in vascular endothelial cells. Biochimie 1988;70:681–686.

20 Hülsmann WC, Dubelaar ML: Carnitine requirement of vascular endothelial and smooth muscle cells in imminent ischemia. Mol Cell Biochem 1992;116:125–129.

21 Giamberadino MA, Dragani L, Valente R, DiLisa F, Saggini R, Vecchiet L: Effects of prolonged L-carnitine administration on delayed muscle pain and CK release after eccentric effort. Int J Sport Med 1996;17:320–324.

22 Kraemer WJ, Volek JS, Dunn-Lewis C: L-Carnitine supplementation: influence upon physiological function. Curr Sports Med Rep 2008;7:218–223.

23 Kraemer WJ, Volek JS, Spiering BA, Vingren JL: L-Carnitine supplementation: a new paradigm for its role in exercise. Monatsh Chem 2005;136:1383–1390.

24 Volek JS, Judelson DA, Silvestre R, Yamamoto LM, Spiering BA, Hatfield DL, Vingren JL, Quann EE, Anderson JM, Maresh CM, Kraemer WJ: Effects of carnitine supplementation on flow-mediated dilation and vascular inflammatory responses to a high fat meal in healthy young adults. Am J Cardiol 2008; 102:1413–1417.

25 Volek JS, Kraemer WJ, Rubin MR, Gomez AL, Ratamess NA, Gaynor P: L-Carnitine L-tartrate supplementation favorably affects markers of recovery from exercise stress. Am J Physiol Endocrinol Metab 2002;282:E474–E482.

26 Ho J-Y, Kraemer WJ, Volek JS, Fragala MS, Thomas GA, Dunn-Lewis C, Coday M, Häkkinen K, Maresh CM: L-Carnitine L-tartrate supplementation favorably affects biochemical markers of recovery from physical exertion in middle-aged men and women. Metabolism 2010;59:1190–1199.

27 Spiering BA, Kraemer WJ, Hatfield DL, Vingren JL, Fragala MS, Ho J-Y, Thomas GA, Häkkinen K, Volek JS: Effects of L-carnitine L-tartrate supplementation on muscle oxygenation responses to resistance exercise. J Strength Cond Res 2008;22: 1130–1135.

28 Kraemer WJ, Volek JS, French DN, Rubin MR, Sharman MJ, Gomez AL, Ratamess NA, Newton RU, Jemiolo B, Craig BW, Häkkinen K: The effects of L-carnitine L-tartrate supplementation on hormonal responses to resistance exercise and recovery. J Strength Cond Res 2003;17:455–462.

29 Inoue K, Yamasaki S, Fushiki T, Okada Y, Sugimoto E: Androgen receptor antagonist suppresses exercise-induced hypertrophy of skeletal muscle. Eur J Appl Physiol Occup Physiol 1994;69:88–91.

30 Kraemer WJ, Spiering BA, Volek JS, Ratamess NA, Sharman MJ, Rubin MR, French DN, Silvestre R, Deschenes MR, Maresh CM: Androgenic responses to resistance exercise: effects of feeding and L-carnitine. Med Sci Sport Exer 2006;38:1288–1296.

31 Spiering BA, Kraemer WJ, Vingren JL, Hatfield DL, Fragala MS, Ho J-Y, Maresh CM, Anderson JM, Volek JS: Responses of criterion variables to different supplemental doses of L-carnitine L-tartrate. J Strength Cond Res 2007;21:259–264.

32 Rubin MR, Volek JS, Gomez AL, Ratamess NA, French DN, Sharman MJ, Kraemer WJ: Safety measures of L-carnitine L-tartrate supplementation in healthy men. J Strength Cond Res 2001;15:486–490.

33 European Food Safety Authority: Opinion of the scientific panel on food additives, flavorings, processing aids, and materials in contact with food (AFC) on request from the Commission related to L-carnitine L-tartrate for use in foods for particular nutritional uses. EFSA J 2003;19:1–13.

Amy Huang
Lonza Ltd
Münchensteinerstrasse 38
CH–4002 Basel (Switzerland)
Tel. +41 61 316 8268, E-Mail amy.huang@lonza.com

Lamprecht M (ed): Acute Topics in Sport Nutrition.
Med Sport Sci. Basel, Karger, 2013, vol 59, pp 143–152

Supplements and Inadvertent Doping – How Big Is the Risk to Athletes?

Catherine Judkins[a] · Peter Prock[b]

[a]HFL Sport Science (part of LGC Group), Fordham, UK, and [b]European Nutraceutical Association, Basel, Switzerland

Abstract

Despite ongoing improvements to regulatory and manufacturing guidelines, the potential for contaminated nutritional supplements to cause a failed doping test for an athlete remains a concern. Several surveys of supplements available through the internet and at retail have confirmed that many are contaminated with steroids and stimulants that are prohibited for use in elite sport. Suggested responses to this issue include the complete avoidance of all supplements. However, this approach seems to be unrealistic as many athletes use nutritional supplements for very different reasons. In addition, the number of publications describing trials that demonstrate the benefit of certain nutritional products has also increased over the last decade or so. This ensures that for many sports the use of supplements will remain a common practice. In response to the issue of contamination in nutritional supplements, many reputable manufacturers have their products rigorously tested by sports anti-doping laboratories to help ensure as far as possible that the risks to an athlete remain minimal. In this chapter we review the issue of supplements and contamination, and look at how this might be addressed through effective quality control procedures at the manufacturing facility and through the highly sensitive testing of finished products using appropriately accredited tests.

Over the last decade there has been considerable scrutiny on the sports nutrition industry following allegations by athletes that positive doping tests were caused by consumption of contaminated supplements. Whilst it is acknowledged that there are a number of supplement products that deliberately contain substances considered prohibited in sport (as defined by the World Anti-Doping Agency list of prohibited substances), research has also shown that a high proportion (anywhere between 15 and 25%) of supplements may inadvertently contain prohibited substances through contamination [1–5]. These substances would not be declared on the label. There is, therefore, a very real risk that an athlete consuming such contaminated products might inadvertently fail a doping test (see later).

This issue demands the attention of a wide range of interested parties, including athletes (from high school to elite) and other supplement users, sports bodies, sports leagues, teams, coaches, dieticians/nutritionists, parents, supplement companies, third party manufacturers (TPMs), national anti-doping organisations, and the World Anti-Doping Agency (WADA).

Athletes will do everything they can to improve their performance. As far as elite athletes are concerned it can be incredibly difficult for them to meet their nutritional requirements through a normal diet. In some cases, energy requirements for athletes in training may be in excess of 5,000 kcal/day – and many will use supplementation as a means of meeting this need. In addition, the high frequency and intensity of training, psychological stress of competition, and travel amongst the athletic population can also make them susceptible to viral infections and some supplements can play a role in managing this problem [6, 7]. Vitamin and mineral supplements can also be necessary to reduce the incidence of injury, cramping, etc. So it is easy to see why athletes may select a range of supplements to ensure they remain in peak physical condition. Of course, there are also many supplements available that do not have proven efficacy, but research behind vitamins, minerals, creatine, β-alanine, electrolyte drinks, milk-based products etc. over the last decade or so has demonstrated that these types of product can be beneficial for the athlete [8–10].

Since the issue of contaminated supplements came to light, many reputable supplement companies have taken extreme steps to ensure that their products are not inadvertently contaminated and are therefore safer for an athlete to use. Such steps include product and manufacturing audits during which the quality controls applied during the production process are scrutinised. These audits are then reinforced through the testing of products for trace amounts of prohibited substances using highly sensitive tests that are accredited to the ISO 17025 standard. Such auditing and testing programmes are typically undertaken by highly specialised sports anti-doping laboratories [11–13].

Inadvertent Contamination

The two main sources of inadvertent contamination (as opposed to deliberate addition) in a supplement are from (1) use of a contaminated ingredient within the product and (2) cross-contamination of products during manufacture.

Ingredients for use in the food and supplement industry are imported from many parts of the world. Some sources of these ingredients may not undertake the necessary quality control measures to ensure that the ingredients they are supplying are not contaminated with prohibited substances. As such, these ingredients may, quite innocently, find their way into supplement products and could result in the final product containing a prohibited substance, which would not be declared on the label.

Similarly, many supplement companies use TPM facilities to produce their products. Such TPMs may manufacture a broad range of products, including those that contain prohibited substances (such as pharmaceutical products, etc.). If the manufacturing equipment is not rigorously cleaned between production runs, it is quite possible that cross-contamination between products can occur at the facility. Examples of the potential for cross-contamination have been shown recently through investigations undertaken at various manufacturing facilities in the USA, Europe and South Africa. In 2011, four separate manufacturing facilities were audited and swab samples were taken from production equipment at each site [HFL Sport Science, unpubl. data]. None of these facilities were known to handle prohibited substances, although all of them handled a wide range of other ingredients that were imported from across the globe. At the two European facilities, 49 and 12% of the swabs taken showed the presence of trace amounts of steroids and/or stimulants. At the US facility, 85% of the 20 swabs taken showed the presence of prohibited substances. The South African facility had a swab contamination rate of 17%. The substances detected included stimulants and anabolic agents and they typically entered the facility through a contaminated ingredient. The presence of these substances on site – even at trace levels – poses a very real risk in terms of cross-contamination into finished products.

Some manufacturers are led to believe that cross-contamination with prohibited substances can be addressed through good manufacturing practice (GMP) certification. Unfortunately this is not the case. There is often a misperception that if a company is GMP certified, the product must have been tested and shown to be 'free' of prohibited substances. Although GMPs are very important, GMP certification will only certify that the manufacturer has the necessary manufacturing documentation, processes and controls in place to ensure product quality, traceability and overall competence. GMP certification does not cover the presence or absence of substances such as prohibited steroids and stimulants at a site. Indeed, it is possible to be GMP certified *and* handle prohibited substances on site. Audits of several GMP-certified facilities in the USA have shown cross-contamination issues with prohibited substances. In one recent case, 32 swab samples were collected from the GMP-accredited facility during a site audit – 15 of these showed evidence for the prohibited substance methylhexanamine [HFL Sport Science, unpubl. data, 2012]. Supplement product samples manufactured at this facility were also analysed and showed evidence for methylhexanamine, demonstrating just how easy it is for contamination on manufacturing equipment to find its way into finished goods.

Which Products Are Most Likely to Be Contaminated?

Unfortunately it is not possible to predict inadvertent contamination. In 2011 the sports anti-doping laboratory HFL Sport Science (HFL) tested 5,009 supplement/ingredient samples at its UK headquarters [HFL Sport Science, unpubl. data]. Table 1

Table 1. Contamination findings in various product types (taken from HFL's analysis of 5,009 supplement samples in 2011)

Product type	Stimulant contamination	Steroid contamination	Diuretic contamination	β-Blocker contamination
Pre-workout products	✓	–	–	–
Post-workout products	✓	–	–	–
Hormone regulators/ antioxidants	✓	✓	–	–
Energy boosters	✓	✓	–	–
Botanical/superfood products	✓	–	–	–
β-Alanine	✓	–	–	–
Weight loss products	✓	✓	–	–
Muscle builders	✓	✓	✓	–
Nitric oxide supplements	✓	–	–	–
Vitamin/mineral products	✓	✓	✓	✓
Testosterone boosters	✓	✓	–	–
Anti-ageing products	–	✓	–	–
Anti-cramp products	–	–	–	✓

shows the contamination findings amongst a number of product categories. There is no apparent trend regarding which kinds of contaminants are found in which kinds of products.

As a general rule, there tends to be a slightly higher incidence of contamination in supplements in capsule and tablet form, particularly if these are made at a TPM [1, 5]. The equipment used to produce capsules and tablets is quite complex and requires very rigorous cleaning procedures if trace cross-contamination between products is to be avoided.

Can Traces of Contamination Really Cause a Failed Doping Test?

Highly sensitive tests are employed when testing supplements for prohibited substances. Generally, detection limits of 10 ng/g, or 10 parts per billion, have been shown to be appropriate to provide adequate reassurance to athletes and manufacturers. This is a *tiny* amount of contamination. Supplement products are therefore tested for the presence of prohibited substances at levels that lie well outside traditional testing of foodstuffs and pharmaceuticals. Generally, impurity testing in the food/pharmaceutical industry is measured at the part per million (ppm) level. So, why is the testing so sensitive for supplements?

The nanogram/gram (or part per billion) detection limits are driven by the fact that the WADA laboratories routinely analyse blood and urine samples with sensitivities in the low nanogram/millilitre region. For example, nandrolone metabolites can be detected in urine at levels <2 ng/ml.

As such, testing sensitivities for supplements also need to be this low. In setting these incredibly sensitive testing limits, WADA is not suggesting that such levels will have any impact on the performance of an individual, but rather that there is no place for e.g. steroids in blood and urine at this level. Finding such a trace amount of a steroid in the blood or urine may be indicative of a much higher dose taken some days previously. So, a urine finding of some steroids at a concentration of 2 ng/ml is enough to prosecute an athlete.

One key issue is to determine the level of contamination in a product that could give rise to such a trace finding in blood and/or urine. A recent administration study involving the nandrolone precursor, 19-norandrostenedione (19-NAS), was undertaken in which 20 volunteers ingested products containing either 1, 2.5 or 5 μg of the steroid precursor [14]. Urine samples were collected from the subjects over a 24-hour period and were analysed for the major metabolite 19-norandrosterone (19-NA) according to WADA regulations for elite athletes. In the study where subjects ingested product contaminated with 5 μg of the 19-NAS, 75% would have failed a doping test had they been drug-tested athletes. For the subjects enrolled on the 2.5-μg product trial, 25% would have failed a doping test. This represents a 'contamination level' in the product of just 0.00005%. A *tiny* amount of contamination, but one that would cause a failed doping test in some individuals. Finally, in the trial where subjects ingested products containing a 1-μg dose of the steroid, 0% of the subjects would have failed a doping test (although 3 subjects did come close to the WADA threshold of 2 ng/ml for the metabolite 19-NA). The authors concluded that for this particular steroid, a dose of somewhere between 1 and 2.5 μg would likely be enough to result in a failed doping test. Figure 1 shows typical excretion profiles for an individual on the three separate trials.

To put this in the context of products – if a product is contaminated with just 10 ng/g (10 ppb) of the nandrolone precursor, and an athlete consumes just 100–250 g (or 100–250 ml) of the product, there is a strong chance that they could fail a doping test. This amount is easily consumed with e.g. whey protein products, electrolyte drinks, etc.

Such a low dose of this substance is highly unlikely to have any performance-enhancing effect for the user, but it could cause the end of an athlete's career.

The supplement matrix can also play a role in how quickly prohibited substances may be detected in the urine following ingestion of contaminated products. One recent study [15] compared the levels of the metabolites 19-NA and 19-noretiocholanolone (19-NE) found in urine after ingestion of a bar and a liquid that had been deliberately 'contaminated' with a known level (10 μg) of the parent steroid 19-NAS. It was found that subjects taking the contaminated bar product had higher urinary concentrations of the metabolites than those taking a contaminated liquid and that, following ingestion, the peak urinary concentrations appeared later compared with the liquid.

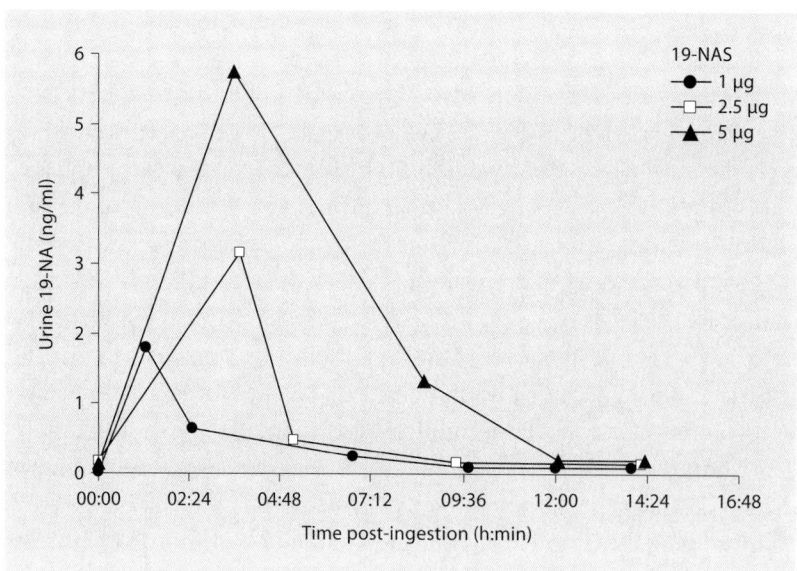

Fig. 1. Typical urinary 19-NA excretion profile for a single subject taking a creatine supplement spiked with 1.0, 2.5 and 5.0 µg of 19-NAS (NB: WADA threshold for 19-NA is 2 ng/ml).

Figure 2 shows the differences in the peak metabolite concentrations in the bar versus the liquid products [15].

What Can Be Done to Minimise Supplement Contamination?

Reputable manufacturers of nutritional supplements can work with experienced sports anti-doping laboratories to have their manufacturing systems audited to identify potential risks from a prohibited substance contamination perspective. In addition to this, products can be tested on a regular basis for trace amounts of prohibited substances, in line with the WADA list [16] using appropriately accredited (ISO 17025) tests. Whilst there is no such thing as a '100% guarantee' that a product is *totally free* of *all* banned substances, this process can significantly help consumers assess the risks when choosing a supplement. The incidence of contamination in products made by manufacturers that have an established history in working within such a quality system is typically less than 1%, showing that this kind of approach can significantly reduce the incidence of having contaminated products [12]. Over time, trends indicate that this figure should reduce even further, showing the long-term effectiveness of such an auditing/testing quality system [12, 13]. Figure 3 shows the decrease in the incidence of contamination over several years for a manufacturer enrolled in a regular supplement testing programme.

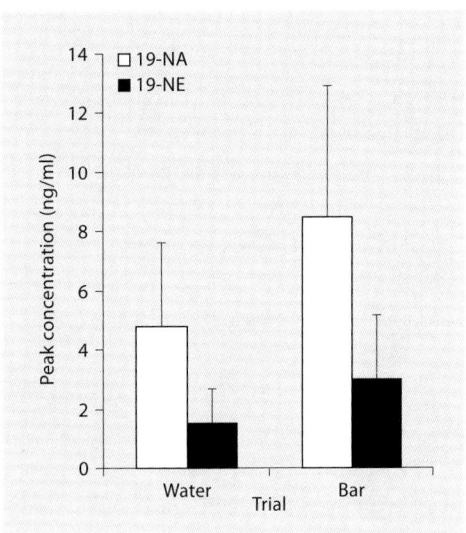

Fig. 2. Peak urinary 19-NA and 19-NE concentrations following ingestion of liquid and bar supplements spiked with 10 µg of 19-NAS.

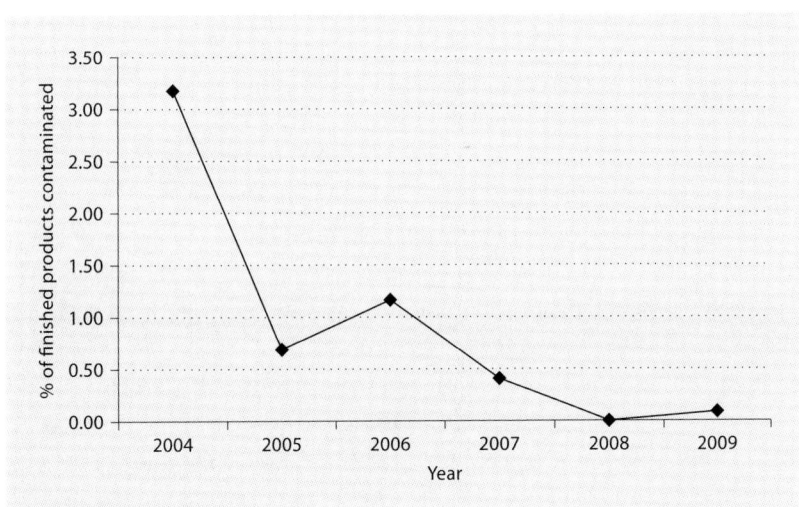

Fig. 3. Reduction in the number of contamination incidences for a typical supplier through working with a regular supplement testing programme.

In 2011, HFL tested 2,702 supplement samples through its 'Informed-Sport' supplement certification programme [HFL Sport Science, unpubl. data]. This quality system includes a rigorous auditing and testing process before allowing products onto the programme and, once registered, products are batch tested for prohibited substances prior to release for sale. In 2011 the incidence of contamination amongst the products on this programme was 0.2% and, most importantly, where contamination was identified, the manufacturer did not release the product for sale. For an athlete, this means that instead of facing a 25% (1 in 4) chance of choosing a contaminated product they can reduce this to well below 1 in 500 if they chose a product that has been through an appropriate certification process. In practice this risk is actually much lower than 1 in 500 as contaminated products are not released for sale, so will not be available for purchase.

As mentioned previously, unfortunately there are no '100% guarantees' with supplement quality assurance systems, but it is clear that they offer an effective risk management tool for the athlete. Some of the key reasons for this lack of 'guarantee' are:

(1) *The Open-Ended Nature of the WADA List of Prohibited Substances.* Each year, WADA will update a published list of substances that are considered to be prohibited within elite sport [16]. Blood and/or urine samples are taken from professional sports people on a regular basis, and these samples are subjected to testing by a limited number of WADA-accredited laboratories across the world. Any evidence for the presence of a prohibited substance will be classed as an adverse analytical finding.

The WADA Prohibited List specifies generic categories of substances (e.g. anabolic androgenic steroids), and then provides a list of specific examples from each category (e.g. testosterone, nandrolone, etc.). The sections on 'anabolic agents' and 'hormones and related substances' finish with a general statement relating to other such substances that are not specified: '. . .other substances with a similar chemical structure or similar biological effect(s)'.

This means that the WADA Prohibited List is not a definitive list of prohibited substances, and these as yet unknown/unnamed compounds that may have a performance-enhancing effect would also be considered prohibited. The most obvious such examples are the so-called 'designer steroids' of which tetrahydrogestrinone was a prime example.

Since no laboratory can test supplements for *'everything'* on the WADA Prohibited List, they cannot offer a 100% guarantee that a product is 100% 'clean'.

(2) *Limitations of Batch Testing.* When testing a supplement product it is only possible to test a small portion of a large production batch. The tests performed on this small portion are highly sensitive and are intended to give an assessment of the likelihood of contamination within that batch. Although manufacturing equipment and blending technologies have much improved over the last few decades, it is still conceivable that variation can exist even within a batch of product. Research has described variation of steroid levels found within the same batch of capsules [17]. Although the cited variation in levels in this paper is rather extreme, this can certainly happen when products are not made to high-quality standards.

It should also be noted that supplement testing services themselves are not regulated by any legislative or sporting body. This can lead to much confusion and concern amongst athletes and manufacturers alike. This lack of regulation can mean that laboratories without the required expertise are able to undertake supplement testing for prohibited substances at the request (and expense!) of the manufacturer. It is therefore vital that anyone seeking to have products tested ensures that they use a laboratory with proven expertise in the field of sports anti-doping and experience of operating within the WADA environment, and that all tests used are accredited to the ISO 17025 standard, with the level of detection (LOD) for each substance in each formulation type (bar, powder, liquid, capsule, etc.) clearly defined. WADA testing of urine and blood samples is also carried out using methods accredited to the ISO 17025 standard. Athletes and their advisors also need to be aware of this issue, and ensure that any products they are using have been through an *appropriate* quality assurance system.

Conclusion

It is clear that there now exists a huge range of supplement products, particularly those available via the internet. It is also very apparent that some of these should be avoided by athletes at all costs if they are to avoid consuming a contaminated (or deliberately adulterated) product. However, there are also many products available that are made by reputable companies, to high-quality standards. Although athletes remain entirely responsible for what they consume, they can be reassured that it is possible to minimise the risks of taking contaminated products. The level of understanding of how supplements become contaminated with trace amounts of prohibited substances has increased vastly over the last decade and reputable companies that enroll in appropriate quality systems to mitigate this problem can offer athletes minimal-risk products.

Disclosure Statement

C. Judkins is an employee and works at HFL Sport Science (part of LGC Group). HFL Sport Science owns and operates supplement testing and certification programmes.

P. Prock sits on the advisory panel for the Informed-Sport supplement certification programme.

References

1 Geyer H, Parr MK, Mareck U, Reinhart U, Schrader Y, Schanzer W: Analysis of non-hormonal nutritional supplements for anabolic-androgenic steroids – results of an international study. Int J Sports Med 2004;25:124–129.

2 Gmeiner G, Hofer H: Untersuchung auf mögliche Verunreinigungen von Nahrungsergänzungsmitteln mit anabolen Steroiden; in Forschungsberichte des Österreichischen Bundesministeriums für soziale Sicherheit und Generationen. Seibersdorf, ARC Seibersdorf Research GmbH, 2002, p 2.

3 Maughan RJ: Contamination of dietary supplements and positive drugs tests in sport. J Sports Sci 2005;23:883–889.

Supplements and Inadvertent Doping – How Big Is the Risk to Athletes?

151

4 Catlin DH, Leder BZ, Ahrens B, Starcevic B, Hatton CK, Green GA, Finkelstein JS: Trace contamination of over-the-counter androstenedione and positive urine test results for a nandrolone metabolite. JAMA 2000;284:2618–2621.

5 HFL Sport Science/LGC Group: www.http://www.lgc.co.uk/divisions/hfl/hfl_sport_science/about_us.aspx (unpubl data).

6 Robson PJ, Bouic PJ, Myburgh KH: Antioxidant supplementation enhances neutrophil oxidative burst in trained runners following prolonged exercise. Int J Sport Nutr Exerc Metab 2003;13:369–381.

7 Peters EM: Exercise, immunology and upper respiratory tract infections. Int J Sports Med 1997;18:S69–S77.

8 Shirreffs SM, Sawka MN, Stone M: Water and electrolyte needs for football training and match-play. J Sports Sci 2006;24:699–707.

9 Benzi G, Ceci A: Creatine as nutritional supplementation and medicinal product. J Sports Med Phys Fitness 2001;41:1–10.

10 Hoffman J, Ratamess N, Kang J, Mangine G, Faigenbaum A, Stout J: Effect of creatine and β-alanine supplementation on performance and endocrine responses in strength/power athletes. Int J Sport Nutr Exerc Metab 2006;16:430–446.

11 Informed-Sport supplement auditing and testing programme: www.Informed-Sport.com.

12 Judkins CMG, Teale P, Hall DJ: The role of banned substance residue analysis in the control of dietary supplement contamination. Drug Test Anal 2010;2:417.

13 De Hon O, Coumans B: The continuing story of nutritional supplements and doping infractions. Br J Sports Med 2007;41:800–805.

14 Watson P, Judkins C, Houghton E, Russell C, Maughan RJ: Urinary nandrolone metabolite detection after ingestion of a nandrolone precursor. Med Sci Sports Exerc 2009;41:766–772.

15 Watson P, Houghton E, Grace PB, Judkins CMG, Dunster P, Maughan RJ: Influence of delivery mode on the urinary excretion of nandrolone metabolites. Med Sci Sports Exerc 2010;42:754–761.

16 World Anti-Doping Agency: www.wada-ama.org.

17 Geyer H, Henze MK, Mareck-Engelke U, Wagner A, Schänzer W: Analysis of 'non-hormonal' nutritional supplements for prohormones, in Schänzer W, Geyer H, Gotzmann A, Mareck-Engelke U (eds): Proceedings of the 19th Cologne Workshop on Dope Analysis, Sport und Buch Strausse, Cologne, 2001, pp 63–72.

Dr. Catherine Judkins
HFL Sport Science/LGC Group
Newmarket Road
Fordham CB7 5WW (UK)
Tel. +44 1638 724400, E-Mail cjudkins@hfl.co.uk

Author Index

Subject Index

intestinal barrier function benefits 64, 65
mechanisms of action 66, 67
Creatine, supplement contamination 147, 148

Doping, *see* Glycerol, Supplement contamination

EGF, *see* Epidermal growth factor
Epidermal growth factor (EGF), bovine colostrum 66

Fruit, vegetable and berry concentrate (FVB)
domains of pros and cons 78–81
exercise performance effects 73–78, 81
health benefits of fruits and vegetables 70, 71
oxidative stress reduction 71–74, 78
FVB, *see* Fruit, vegetable and berry concentrate

GABA, *see* γ-Aminobutyric acid
Gastric emptying, milk protein effects 124, 125
GH, *see* Growth hormone
β-Glucan, *see* Pleuran, Upper respiratory tract infection
Glycerol
exercise performance benefits 110
hyperhydration
dose 108
duration of fluid retention 109
fluid retention promotion 104
fluids
timing 109
types 109
volume 108, 109
negative effects 106, 107
prospects for sports use 111
rehydration
efficacy 109, 110
regimens 107, 108
side effects 110
thermoregulatory and cardiovascular benefits 110
World Anti-Doping Agency ruling 105, 106
Glycogen, low-fat chocolate milk consumption response 129, 130
Growth hormone (GH)
γ-aminobutyric acid
hormone secretion induction 40, 41

supplementation for hormone secretion induction 41–44
functions 36
release regulation 37

Hyperhydration, *see* Glycerol

Immune function, *see also* Upper respiratory tract infection
bovine colostrum benefits 65, 66
β-glucans in protection 58–60
strenuous exercise and immune suppression 57, 58, 63
Inflammation
cherry juice anti-inflammatory activity 88–90
cytokines 87
Intestinal barrier function
bovine colostrum supplementation effects 64, 65
components 48–50
exercise effects 50
leaky gut 47, 48
probiotic supplementation effects in exercise 52–54
surrogate markers 51, 52

Leaky gut, *see* Intestinal barrier function
Leucine, milk consumption and body composition response 95, 100
Lipopolysaccharide (LPS), intestinal barrier function marker in plasma 51, 52
LPS, *see* Lipopolysaccharide

Milk consumption
bioactive components 95
body composition response 95
low-fat chocolate milk
chronic supplementation studies 131, 132
composition 128
exercise recovery 128, 129
muscle glycogen response 129, 130
muscle protein synthesis 130, 131
prospects for study 100, 101
rehydration after exercise
milk protein effects 122–124
overview 121, 122
prospects for study 124, 125
resistance exercise studies in women
acute response 96

bone health 98, 99
chronic studies 96–98
weight loss 99, 100
Muscle
carnosine, *see* Carnosine
cherry juice for injury and recovery
benefits 90, 91
low-fat chocolate milk consumption
glycogen response 129, 130
protein synthesis 130, 131

Nitrate, dietary
exercise performance effects 32, 33
mechanism of physiological effects 31–33
oxygen uptake relationship 30
prospects for supplementation studies 33,
34
Nitric oxide (NO)
contamination of supplements 145
formation from reduction of nitrates 29, 30
supplements for enhancement, *see* Arginine,
Citrulline, Nitrate, dietary
synthesis 18, 19, 23
NO, *see* Nitric oxide

Osteoarthritis (OA), cherry juice anti-
inflammatory activity 89, 90
Oxidative stress
cherry juice antioxidant capacity 87, 88
fruit, vegetable and berry concentrate
reduction 71–74, 78

Plasma volume, salt loading effects before
exercise 116
Pleuran, upper respiratory tract infection
protection studies 59, 60
Probiotics, intestinal barrier function effects in
exercise 52–54

Reactive oxygen species, *see* Oxidative stress
Rehydration, *see* Glycerol, Milk consumption
Resistance exercise, *see* Milk consumption

Salt loading
delivery after dehydration but prior to
exercise 114, 115
delivery during dehydrating exercise 115,
116
effects before exercise
plasma volume 116
exercise performance effects 116–118
Supplement contamination
doping test failure risks 146–148
good manufacturing practice 144
minimization 148–151
scope of problem 143, 144
susceptible products 144, 145

Testosterone
carnitine in androgen response 139, 140
contamination of booster supplements 145

Upper respiratory tract infection (URTI)
bovine colostrum supplementation
studies 65, 68
β-glucans in protection 58, 59
pleuran supplementation studies 59, 60
strenuous exercise and immune
suppression 57, 58, 63
URTI, *see* Upper respiratory tract infection

Vegetables, *see* Fruit, vegetable and berry
concentrate

Whey protein, *see* Milk consumption
World Anti-Doping Agency (WADA)
glycerol ruling 105, 106
list of prohibited substances 150
supplement contamination and doping test
failure risks 146–148
testing sensitivity 146, 147

Zonulin, intestinal barrier function marker 51